J. J Armistead

An Angler's Paradise

and how to obtain it

J. J Armistead

An Angler's Paradise
and how to obtain it

ISBN/EAN: 9783741107610

Manufactured in Europe, USA, Canada, Australia, Japa

Cover: Foto ©berggeist007 / pixelio.de

Manufactured and distributed by brebook publishing software
(www.brebook.com)

J. J Armistead

An Angler's Paradise

AN ANGLER'S PARADISE

AND

HOW TO OBTAIN IT

——— ———

J. J. ARMISTEAD.

THE AUTHOR.

AN ANGLER'S PARADISE

AND

·HOW TO OBTAIN IT.

J. J. ARMISTEAD

Author of "A SHORT HISTORY OF PISCICULTURE,"
Lecturer on "FISHCULTURE," "IMPROVEMENT OF FISHERIES," ETC
Member of Royal Commission on Tweed and Solway Fisheries,
Proprietor of the Solway Fish Farm, School of
Fishculture, Etc , Etc.

THIRD EDITION.

THE ANGLER, AT THE OFFICES, SCARBOROUGH.

1898.

PRINTED BY
THE GAZETTE COMPANY, LIMITED, 31, ST. NICHOLAS STREET, SCARBOROUGH.

TO

SIR HERBERT EUSTACE MAXWELL, BART., M P.,

THE FOLLOWING PAGES

ARE RESPECTFULLY DEDICATED

BY

THE AUTHOR

PREFACE.

I HAVE frequently been pressed to write a book on fish culture, and at the earnest request of many who take a great interest in the subject I undertook the task, and this volume is the result. I trust much information may be found in its pages that will prove useful to many of those who read it. I know what a help such a book would have been to me when I first began to study the subject.

I have divided it into two parts. In the first four chapters I have made mention of some well-known angling resorts, of the Solway Fishery, and of the progress of fish culture ; and for the second part of the work I have written some thoroughly practical chapters, which are intended to serve as a guide to those who are the fortunate possessors of suitable water, and which will enable them to develop the resources which have so long lain dormant.

If the style in which I have written be somewhat varied, it must be borne in mind that the
conditions under which the writing has been done
have also been varied, and this must be my apology
for any deficiency which may be found in its pages.
Most of the chapters have been written off without
the opportunity of reference to other books. Several
of them have been penned whilst crossing the
Atlantic, some in railway station waiting-rooms,
often during a midnight wait, or in the early hours
of the morning. A part was written at sea in the
cabin of a trawler, and the rest has been put
together amidst the scenes of a very busy life. For
my knowledge of the subject I am indebted largely to
my own perseverance. and determination to understand thoroughly that which I had taken up as my
life's work, as well as to many fellow-labourers in
the same field, who have given me the benefit of
their knowledge.

It is interesting to look back upon the work of
thirty years that have passed away, and to note how
one has been led. and how the work has developed,
at times under considerable difficulties, until it has
reached its present magnitude. The present seems
a fitting opportunity for tendering my warmest
thanks to some of those who have aided me in my
investigations of so fascinating a subject. To the
late Frank Buckland I am indebted for my first

introduction to the study, and to the late Dr. Francis Day for his kindly help on various occasions when it was my privilege to meet him. But for the kindly advice and encouragement of the former the Solway Fishery would never have been commenced.

To my friend, George H. Brocklehurst, B.Sc., I am indebted for a considerable amount of assistance in microscopic research and the study of embryology, etc. In studying the mysteries of our ponds and ditches, and working out the life histories of many of the creatures contained therein, I have been largely assisted by my friend, Leonard West, Esq., of Darlington, who has kindly supplied the drawings in illustration of the chapter on pond life, as well as a considerable portion of the letterpress. To Mr. Thomas Bolton, of Birmingham, I am also indebted for information respecting the microscopic inhabitants of our waters, and also for drawings illustrative of the embryonic life of a salmonoid.

For views of the Solway Fishery I am indebted to Mr. Willie Anderson, of Partick, Glasgow, and to J. Rutherford, Esq., of Jardington, Dumfries; and I am glad to take this opportunity of acknowledging the kindness and hospitality of many who will read these pages, who have, from time to time, permitted free access to their waters for the purposes of scientific research, etc. Without the help of some of these this book would in all probability never have seen the daylight, and many pleasant memories are recalled of excursions by river, brook, and mere, by the writing of its pages.

J. J. ARMISTEAD.

Solway Fishery, Dumfries,
December, 1894.

PREFACE TO THE THIRD EDITION.

THREE years having almost elapsed since the publication of the Second Edition, which is now nearly cleared out, and the demand still keeping up, I have decided on issuing a third. The price will remain the same until half the Edition is sold, when it will be raised to 12/6. This is rendered necessary owing to the number published being smaller.

J. J. ARMISTEAD.

Solway Fishery, Dumfries,
November, 1898.

CONTENTS.

ILLUSTRATIONS.

Part 1.

AN ANGLER'S PARADISE

AN ANGLER'S PARADISE.

CHAPTER I.

INTRODUCTORY.

Referring to what has been done at home and abroad—In New Zealand—In Tasmania—Taking Salmon by machinery in America, etc.

YES! Trout culture as a means of stocking waters is a success in Britain. Its benefits are already being largely felt in many quarters, and it only requires to be more widely known and its advantages understood, and it will be extensively taken up by those who have the necessary facilities for availing themselves of its benefits. Its success is proved beyond doubt, by the results which have accrued of late years to the stocking or restocking of waters, when judiciously done. I say judiciously done, for much depends on this.

I am aware that many sweeping assertions have been made to the contrary, but results have proved them to be incorrect. It is true that in bringing the matter to the successful issue that it has reached there have been many failures, and it is owing in a measure to some of these failures, that the results are not infinitely larger to-day than is in reality the case. They are of two descriptions :—

1. In many instances in which fish culture has been attempted, it has been by persons who have carried it to a certain point, where it has ended in failure or produced no appreciable result, and so it has been abandoned and frequently a bad name has been given to it. "No result has accrued," they say, "after

considerable outlay in stocking waters." It is here that the mistake has been made, through mismanagement, for the waters referred to have probably not been stocked at all, as stocking is understood to-day. Possibly a quantity of unhealthy or badly grown fry have been turned out to die, or, it may be, have been killed by the operation, or turned out in places which were totally unsuited for them, and, as might be expected, no improvement has followed. Such failures as these have undeservedly brought discredit on the work, and they have unfortunately had the effect of retarding its progress.

2. The time was but a few years ago when even fish culturists of experience, who were looked up to by others as such, were frequently not rewarded by successful results to their many experiments.

Upon these latter failures, however, is now built up a mass of knowledge which has enabled us to bridge over the many difficulties of trout culture, and to overcome what at one time appeared to be insuperable barriers to the further progress of this interesting study. The tendency of late years has been for a great reduction in the prices of our products, but all this time fish as an article of food has been getting dearer. The cause is obvious, and the remedy is obvious, and the sooner it is applied the better. Do with fish what is done with cattle and with poultry, and the benefit accruing will be very great. Every country house with a water supply should have its fish ponds, for the purpose of supplying the kitchen as well as for angling. This work is now being taken up, and where it is properly carried out the results are highly encouraging. Existing ponds have been utilized most advantageously, and in other cases new ponds have been made on approved principles, and the success of the latter, where properly managed, has been very great indeed.

By way of illustration I may mention a pond which I constructed, which measured about ninety feet by thirty and averaged about five feet in depth. Out of this pond was taken over seven hundred pounds weight of trout. These fish had occupied the pond for three years, and for a year previous to their introduction a very small one had contained them. Now multiply the area of the pond by fifteen and the result is four thousand five hundred square yards, or less than an acre. An acre of such water, then, would produce at the same rate over ten thousand five hundred pounds of trout in three or four years. Taking the value of this at a shilling per pound the result is £525. Experience teaches us that the cost of producing and maintaining these fish could be brought under £325, which leaves a margin of £200, or at the rate of £50 per annum per acre—a very handsome profit indeed. Of course there is risk, and results would vary, but against this may be placed the fact that I have taken a decidedly low estimate. I have indeed simply stated what has been done and not what may be done.

It is now more than a quarter of a century since I began my fish-cultural operations, and during that time I have had, in common with all fish culturists, many failures. I think I may fairly say, however, that from every one of these I have learned something, and often that something has been well worth the cost that it has been to me, even though it involved the loss of a considerable number of fish. Losses are always grievous at the time they occur, but the knowledge which we now possess may be said to have been largely gained, or to have grown, out. of these misfortunes.

It is not to be expected that everyone who attempts to manage a fish farm will carry his work to a successful issue; indeed it is only in the hands of a skilful operator that such a work can result in eventual success. Everyone who knows anything of cattle or agriculture, is well aware that it requires a good technical and practical education to make a farmer. So it is with the breeding of fish, but it is as certain in its results when in proper hands as the breeding of cattle, poultry, or anything else, and quite as easily managed as it becomes understood. It however, requires, as I have said, a considerable amount of education, and it cannot for one moment be expected that it can be carried on at once in a successful manner by everyone who inclines to take it up. Education must be had in one way or another. Either beginners must pay for instruction at some well-conducted fish-cultural establishment, or they must flounder in the mire for some time, it may be a considerable time, and patiently bear many losses, discomfitures, and failures, before they can hope to bring to a successful issue that which has taken many years of patient study and toil to establish, in the case of those now successfully carrying on fish farms in this and other countries. I say this most emphatically lest anyone should be deceived.

But there is a most important and valuable branch of fish culture which ought to be carried on by most if not all of those who possess the facilities for it, and this is the growing of the fish themselves after they have been reared on a fish farm. This is a work in which anyone possessed of ordinary intelligence can advantageously engage, after having studied the subject a little ; and my chief object will be to show how a large percentage of the waters running waste in this country may be utilized, and great benefit derived from their successful cultivation. The work is now being done, and can be done again in hundreds of other places. Fish culture is nothing new after all, but there is no need to repeat at length its ancient and modern history here. I referred to that subject in a pamphlet I published more than twenty-five years ago, and it has been well traced out by many other writers. It will suffice, therefore, to say that fish culture was known to, and carried on extensively by, the Ancients ; and even in later times our abbeys and monasteries possessed extensive fish ponds, traces of which remain in fairly good preservation to the present day. A few which I have inspected may perhaps some day be again put to their proper use ; they might then pay a dividend. Fish culture is successfully carried on in China, and has been, I believe, from time immemorial.

It was commenced in New Zealand over twenty-five years ago,* but on a very small scale indeed at first, eight hundred trout ova being successfully hatched in the year 1868, which were obtained from the natural spawning beds in Tasmania. Now we find that the first introduction of trout into Tasmania was effected in 1864, a small number of eggs being sent out from this country by Mr. Frank Buckland, Mr. (now Sir J. A.) Youl, and Mr. Francis

---

*"I am reminded by the Editor of the *Field* that the acclimatization of *Salmonidæ* in Tasmania was discussed as far back as 1841 by the Colonists themselves, and it was through the unwearying efforts of Sir J. A. Youl that they were first introduced into Antipodean streams. The names which should be always first recognised in connection with the work are those of Mr. Ramsbottom, of Clitheroe, and Mr. (now Sir Thomas) Brady, and Mr. Edward Wilson, president of the Victorian Acclimatization Society."

Francis, the number being about two thousand seven hundred altogether. As a result of the importation of trout ova into Tasmania and their cultivation there, we find in four years that country sending ova, taken from fish on the natural spawning beds, to New Zealand. We find also that those eggs were successfully hatched there, and from this small stock a beginning was made, and there seems to be little doubt that from these eggs trout originated in New Zealand. So successful was the work carried on there, that the New Zealand Government very wisely took it in hand, and the result was a considerable importation of ova into the colony, the Solway Fishery having had the honour of furnishing some of these. It is largely owing to the work of the Acclimatization Societies, however, that the fish-cultural work so wisely fostered by the Government has prospered.

In order to shew the difficulties under which it was carried on in the early days of its history, I quote the following from the *Lyttelton Times*, N.Z., regarding a consignment of ova received in 1873, and shipped by Mr. Frank Buckland :—

" The first of the consignment for Otago proved to be entirely bad when opened, and the second was very little better. But very few fish indeed were hatched out in Otago, and after being liberated in the rivers nothing more was heard of them, whilst Mr. Johnson failed to bring any of the five or six thousand presented to this province to life. The ova on this occasion were obtained to the order of the General Government, from the Stormontfield hatching establishment on the Tay, Scotland. Having been placed in boxes containing a few hundred each, they were conveyed to London; each parcel was supported the whole distance by hand in order to prevent jar. An icehouse for the reception of the ova had been constructed in the forehold of the vessel, and about twenty tons of ice were used in packing around the boxes. The voyage occupied a hundred days. The curators of the Canterbury and Invercargill Society lost no time in going on board, where they succeeded in making arrangements for getting the ova out on the following day. The Invercargill ova were placed in a large case and covered with ice and straw, whilst cases about four feet by three feet, and three feet deep, had been made for the Canterbury portion of the shipment, each case being

lined with zinc, over which was a coating of flannel to prevent the
ice melting.

"The ova boxes were fixed in tight with horse-hair and ice,
added to which the cases containing the boxes were covered on
the outside with matting, so as to resist the power of the sun.
Mr. Johnson also took care to do his packing as near the ice-house
as possible. The Southland ova were despatched by a steamer
specially chartered, while the cases containing the Canterbury
boxes of ova were placed on board the s.s. *Alhambra*, suspended
by indiarubber slings in order to prevent jar of any kind. The
Alhambra sailed from Port Chalmers at four p.m. on Monday,
May 5th, and arrived at Lyttelton about the same hour on Tues-
day, May 6th. Mr. Johnson had the ova conveyed to the
Lyttelton station, where a special train was in waiting, and the
boxes were suspended by indiarubber slings. The train was only
driven through to Christchurch at a slow walking pace, and the
boxes were left at the station during the night.

"When the ova boxes were opened there was found to be a
great difference in the condition of the contents, the whole of the
ova in some being entirely bad, while in others there was a large
percentage with a healthy appearance. The first layer of moss
having been removed from these boxes, the ova were emptied into
a stream of running water, which had been previously iced, and
subsequently all ova shewing the slightest signs of life were taken
out of the stream, and placed in the hatching boxes in the fish
house. It may be said there were from one thousand to two
thousand placed in the boxes, and that there are several hundreds
of these which have a very promising appearance."

With regard to the same shipment, the following is an extract
from a letter published in *Land and Water*, of August 2nd,
1873, and received from Mr. Henry Howard, dated Wallacetown
Trout Ponds, N.Z., 14th May of that year:—

"On arrival at the ponds and opening the boxes, I found the
temperature in the moss was 43°, and as our springs are always at
or about 50°, I reduced the water to the former temperature by
putting ice in the upper cross-box. Water also reduced to 43°
was put in large pans, and the boxes sunk in them so as to leave
all the ova submerged. The ova were carefully separated with

spoons and placed in the hatching boxes; it would have been impossible to turn out the ova as you suggested, as nearly all stick more or less to the moss, and I must have put in masses as large as one's hand, of dead and living ova mixed. The boxes varied greatly. In one I counted over three hundred good, and I only took sixteen good from one lot of three boxes. It was quite easy to predict which were good the moment the lid was removed. The good ones had a green healthy look under the moss, and no fog under the lid; the bad ones had a fog, not unlike what I remember gossamer webs to be in the autumn mornings with a heavy dew; the moss also was browner and sunk in the boxes, whereas the good boxes were light but full.

"I believe the kind of moss has much to do with it. The brown moss had a good deal of old grass and sticks amongst it, as if taken from woods; but the good was more like what grows on the boles of trees and about sluice gates. I could see no sign of eyes in most of the ova, but in some the fish were plainly to be observed. I feared therefore at first, for a few days, that many were unfertile, but this morning I see the eyes in many more, and the deaths are far less. No doubt the warmer temperature is beginning to tell. You will be anxious to know how many are likely to survive. I can only give a rude guess, but from the look I should think there are not more than from fifteen to twenty thousand left, but many of these will of course come to nothing; if we hatch from six to ten thousand I shall consider we have done well."

It is plain from these extracts that in 1873 the work in New Zealand was being carried on under considerable difficulties. Let us now look at the state of things out there ten years later (1883), and we find that trout are thoroughly acclimatized, and are being cultivated in many places. Amongst other cases is one mentioned in the *Otago Daily Times* about this time, which gives the following account of Mr. W. S. Pillans, of Otago, who had successfully raised some six thousand trout of his own :—

"This gentleman's property, known as Manuka Island Station, is situated a few miles from Balclutha, near the bank of the river. It boasts no natural advantages for pisciculture, and what has been done has been done by hard and persevering work. The number quoted (six thousand) does not, however, by any means, sum up the extent of Mr. Pillans' operations so far.   He has, it is said, given a quantity of fry to the Acclimatization Society, and exclusive of these, he has now at his nursery fifty-seven thousand ova in course of hatching, six thousand yearlings, and nearly two hundred and fifty two-year-old fish, all apparently thriving exceedingly well in the limited space at their disposal."   The report goes on to describe the "stripping" of the trout, and manipulation of the ova and fish at Mr. Pillans' hatchery, of which it gives a very interesting account.

In looking through the " Report of the Otago Acclimatization Society for 1891 " I see a most gratifying feature, and that is a balance-sheet of the Society showing a profit over and above working expenses, resulting from the sale of licenses, sales of trout, etc.   The Society began the year with a balance in hand on fish-culture account of £248, and at the close of the year, after paying all expenses in connection with the work, the account shews a balance of £396, or about £148 profit on the year's working. When we take into consideration all that had to be done, and that the working expenses were excessive, this result speaks volumes.

Now let us take the state of things in New Zealand to-day, and what do we find?   Why, that the rivers of that country are many of them full of magnificent trout, that have grown beyond all expectation.   Trout cultivation in New Zealand is a grand success.   The attention bestowed on the fish by the Wellington Acclimatization Society and the Otago Acclimatization Society is most praiseworthy, and has produced the most gratifying results, and much larger fish than the average in Britain are very plentiful. In the records received from time to time I see such cases as thirteen fish, a hundred and two pounds ; about eight pounds each on the average.   Fish from five to ten pounds are common, and trout varying from thirty to forty pounds are reported as having been occasionally taken.   Take, as an example, one out of many reports which have appeared in our papers, the following from *Rod and Gun*, March 11th, 1893 :—

"The Rakaia is a river of forty or fifty miles, stocked with the finest trout, seldom under two pounds, and not uncommonly containing weights of twelve pounds, fifteen pounds, and even twenty pounds, and it is confined ordinarily to three or four large streams, breaking up constantly into numerous smaller ones. It is stated that the Rakaia is the home of some of the finest trout in the world, and that fish get larger and more plentiful the nearer it approaches the sea. They are often caught in the salt water at the mouth of the river itself. It is one of the snow-fed rivers; and with respect to it, and all streams that so take their rise, the sensible warning is given that the angler should always make inquiry as to the state of the water before proceeding upon an expedition. It was in Rakaia that the splendid takes of trout reported last year were made. Eight fish, weighing eighty-eight pounds, is something indeed worth entering in the diary; but that was beaten in February the year before last, by one angler taking thirteen fish weighing a hundred and forty-seven pounds, the largest being sixteen and a half pounds, the smallest eight pounds. Even this, however, was surpassed by a party of three gentlemen fishing along the shingle bank, on the south side of the river, near to the sea. They took forty-four fish weighing three hundred and forty-seven and a half pounds." Such reports as these have been numerous and are most encouraging.

From Tasmania also come most gratifying accounts, as well as from other countries, and a friend writing from Tasmania remarks :—"The English trout that have been acclimatized here have done remarkably well and attain a great size." So then in Tasmania also, trout culture, though carried on under the great difficulty of importing ova from Britain at a time when the matter was but very imperfectly understood, has proved a decided success.

The results obtained in America would fill a series of books, and want of space must be my plea for not going into details respecting them. I may say that I have seen quite enough, when over there on fishery business, to be convinced that a great work is going on, both in the United States and in Canada, that is productive of much good. Salmon have for some time been caught on some of the great rivers by machinery, large wheels being fixed at suitable places in the streams, which literally scoop or shovel the fish out by thousands. So enormous is the destruction of salmon on some of these rivers that there would 'soon be none left but for cultivation. The fish crowd up the rivers in the migratory seasons in enormous numbers; quantities, in fact, that we have no idea of in this country. The wheels are placed on scows or barges, or worked from the side of the river, just as may be most convenient for taking the fish. These latter in ascending rivers follow the main currents, and an expert is able at once to fit up a wheel that will do execution among them. The apparatus has, roughly speaking, some resemblance to a large water-wheel fitted with big skeleton scoops covered with netting. The fish in their ascent of the rivers swim into these, which revolve in the opposite direction, and they are carried up to the top of the wheel, when they drop through a shoot, which sends them into a receptacle alongside or behind the machine as the case may be.

An attendant knocks them on the head, strings a lot of them together by means of a rope, which is then fastened to a ring in a barrel and the lot flung into the stream. So the work goes on, and these strings of fish are carried down stream for some distance, when they are picked by a small steamer on the look-out for them and taken to the canneries. These wheels are notably in use on the Columbia, Clackamas, and other rivers.

Concerning the cultivation of the salmon on some of these rivers, the eminent American fish culturist, Livingstone Stone, says, in one of his letters :—" . . . In regard to the success that has attended the culture of the *Salmonidæ*, the Government station for hatching salmon on the McCloud river, California, may be mentioned as an unquestionable instance of labour in that direction well rewarded. It is universally acknowledged that the hatching of salmon at this station, which I had the honour of naming Baird after our distinguished Commissioner, has immensely increased the number of salmon in the Sacramento River, of which the McCloud is a tributary. . . . The good effect of the hatching of salmon at the Government station on the Clackamas river in Oregon, is doubtless very similar. Although the limited output of young salmon at this station is wholly inadequate to the demands of so great a river as the Columbia, of which the Clackamas is a tributary, nevertheless the salmon production, such as it is at this station, is believed to be of immense benefit to the river, and it is thought to be almost certain, that, without the help of the Clackamas hatchery, the enormous drain on the salmon supply of the river made by its numerous canneries would have caused an alarming diminution of the salmon of this wonderful salmon river. I think it is safe to say further, that unless the hatching and distribution of young salmon is continued at these stations, either the canneries on these rivers or the salmon in them will become a thing of the past.   " Very truly yours,

           " LIVINGSTONE STONE."

The result of acclimatizing and cultivating various members of the *Salmonidæ*, as well as other fish, on some parts of the Continent, notably in Germany, has been decidedly successful. Trout are now grown there in large numbers, both for the market and also for stocking waters, and their angling localities are already being advertised in British newspapers. But in our little islands strides are also being made in fish-cultural departures, as I shall endeavour to shew in my next chapter.

CHAPTER II.

Having reference to the Solway Fishery—Loch Kinder and its trout—Loch Leven trout—An angler's paradise—Poachers—Nature's motive power.

"AN angler's paradise!" Three words that are at once suggestive to the lover of the rod of a wonderfully pleasing sensation. And such were the words that escaped the lips of one of the pleasantest and most agreeable anglers who ever visited the Solway Fishery. It came about in this way. I had been down the coast yachting for a few days, studying some of the denizens of the deep sea with a view to their cultivation, and on my return found a card had been left by a visitor who had called to see my fish ponds, and who was described to me as taking great interest in the work, and who, after seeing part of what was to be seen had exclaimed, "What an angler's paradise!"

It struck me as being a most refined and appropriate expression, in few words, of the feelings of one who afterwards proved himself to be one of the best anglers I had ever come across. He always caught fish, and what was more, he got them when nobody else could catch them, and in the most skilful and sportsmanlike manner. On loch or stream it was the same. One day I lent him my boat on a loch where other people seemed to have only indifferent success, and he soon captured over forty pounds of pike and perch, some of the latter being about three pounds in weight.

It is rather amusing to note the exclamations of visitors on being shown round my fish ponds, expressive possibly of pretty much the same kind of feeling, but certainly varying a good deal in phraseology. My friend to whom I have just referred gave vent to his feelings once and for all in three very expressive words. Another interested party kept on continually uttering the words

"By Jove," and at nearly every pond we came to this expression was repeated. Considering that there are upwards of seventy ponds, and he examined most of them, he must have made use of this expression over forty times during his round. But he exhibited, although not so demonstratively, the same keen inward sense of enjoyment, and the repeated utterance of these two words proved no doubt a relief to him. Another visitor kept continually exclaiming "Dear me," and the number of times those two words were repeated must have been something considerable, whilst yet another gave vent to his feelings by saying "What a caution!"

Most visitors have expressed great delight at the sights shown, and have evidently highly enjoyed their visits, whilst yet a few would gaze complacently on the scene as if it were all a matter of course, and as if they had seen the same thing a hundred times before. Need I say these were not anglers? No feelings were aroused within them, and no interested enthusiasm was observable as they gazed upon the masses of fish, which to them no doubt were no more suggestive than a pile of herrings on a costermonger's barrow.

> "Breathes there a man with soul so dead,
> Who never to himself hath said,"

nor to anyone else either, anything which would be indicative of the fact that he was undergoing that deep sense of pleasure which every angler experiences at even a far less impressive sight.

I hope that no reader will for a moment think I am unduly poking fun at him—far be it from me to do that; but whilst giving an indescribable amount of pleasure to so many who come here, it is only fair that we who dwell in this most delightful wilderness should get some amusement in return. I am sure all will agree to this.

But to return to my first friend; I found he was to be in the neighbourhood for a couple of months, having taken a house along with the fishing on a good trout loch and the stream connected with it, so took the earliest opportunity of driving over to make a call. He had just come off Loch Kinder with about fifteen pounds of trout, which were brought in for our inspection. Beautiful creatures they were and in excellent condition, averaging about or near a pound each. After a very pleasant fishy chat, and an inspection of his rods and tackle, which were "something numerous" and formed an enticingly interesting collection, and promising, all being well, to meet again another day, we bade adieu.

But more about Loch Kinder and its trout. The loch has long been famous for them, and tradition tells us they were originally brought from Loch Leven by the monks of the abbey close by. Be this as it may, the fish are excellent, and some of them which I have seen bear a resemblance to some forms of *Salmo Levenensis.* But this latter fish is found to occur in great variety, and according to the testimony of Dr. Francis Day and others, it soon assumes in some waters an ordinary *fario* type, shewing more or less of red on the adipose fin and having red spots. I have made a careful study of this trait in the character of the Loch Leven trout, and I find that even if fish are bred from a pair of typical Loch Levens, or, as we call them on a fish farm, "thoroughbreds," that some, it may be only a few, show these variations, which are looked upon as typical of the ordinary *fario* form of trout. It is only by careful study that these peculiarities can be followed up, but those who are accustomed to handle, year after year, large numbers of these fish of all ages and sizes have an opportunity of noting changes and differences which few others possess.

Much has been done but much yet remains to be done in tracing the development of new types, the result it may be of artificial cultivation, and in tracing out the reversion in the case of some individuals, to an origin from which they may possibly have sprung. It is not my province to go into these matters here, but I cannot pass by a subject of such deeply fascinating interest without a brief allusion to it.

Wherever these Loch Kinder trout originally came from, they are good fish. There seem to be three varieties in the loch, due no doubt to the different portions the fish inhabit, a feature noticeable in many other lochs. The weakest point about this piece of water is its great want of accessible spawning ground. At present it is entirely inadequate to the requirements of the loch, and beset with natural difficulties, which if removed would make it one of the finest lochs in the south of Scotland, and the work could be quite easily done.

Most of the trout are under the necessity of passing down stream for spawning purposes, and this they do until they reach the junction of the stream with another known as the Glen burn, when they alter their course and head up the latter in large numbers, pushing right up to its very source, as well as up some of its tributaries. This description of Loch Kinder is applicable, in a more or less modified degree, to a great many of our lakes, the fishing of which might be very materially improved by a little judicious interference on the part of man. I have met many an angler who has very pleasant reminiscences of days spent on the loch, but all agree in saying, that it ought to yield them a far better basket of fish. And so it would under slightly altered conditions.

If it be possible to keep up such a constant supply of trout as is now being done in Loch Leven, notwithstanding the pike, how much more easily it might be done in other lochs where no pike exist. Loch Awe, too, is another instance, and I might mention more. Even in lakes containing pike it has been proved that, by using proper means, a splendid stock of trout may be maintained, and it is idle to talk longer of the uselessness of dealing with such waters. That day is now gone by.

There was a time when here the land was full of magpies, jays, and hawks ; more of these birds than grouse and partridges. How now ? I have hardly seen a magpie or a jay the last ten years in this part. Hawks there are a few, because they're migratory. And also as to mammals ; why, the polecat and the badger are about extinct, the wild cat's gone, the stoat and weasel too are very scarce, but grouse and rabbits are abundant where formerly they were rarely seen. Pike can be killed or caught alive, and pike are valuable in their proper places, and they should not spoil but help to make "an angler's paradise." So let us put them right. These little matters will get shaken down after a while.

I have seen many lakes in which a stock of trout might easily be maintained notwithstanding the pike—waters that now are comparatively barren—the existing pike being chiefly small, having little to feed on but each other, and perhaps a few perch. A few years ago it was supposed that the presence of pike in any water was an insuperable difficulty in the way of stocking with

C

trout. This is not so, as we have seen, and I fully believe that before long it will be plainly demonstrated that pike can be sufficiently banished, when desired, from a great many of our waters in which they now exist. When the mode of dealing with them is as well studied and practised as that of dealing with land pests, we shall soon find out ways and means for getting rid of them.

There is an old saying, that he who makes two blades of grass to grow where only one grew before, is a benefactor to his fellow-men. Many people have dealt with blades of grass, and as we are well aware, some of them have met with a considerable amount of success. Now there are a few of us who have devoted our energies to something else, and I hope to show that it is not only quite possible, but often comparatively easy in a great many cases, to make fish grow where but few fish grew before. The fish culturist has been a long time in " coming to the scratch," but take care he does not beat the botanist after all. That this growing of fish is not only possible, but has in many places already been done successfully, is now an established fact.

Ponds have recently been constructed in many places by those who have become alive to the advantages of fish culture as it is carried on at the present day. I have seen as a result the delighted angler filling his basket, not with little fingerlings, such as we have been accustomed to catch in so many of our mountain streams, but bringing to bank, after tough resistance, fish after fish, requiring the use of the landing-net, and weighing pounds instead of ounces.

To leave the busy din and bustle of the city, and after a comfortable journey, as it is now accomplished by rail, to find one's-self located in the

" Land of the mountain and the flood,"

to see the mists creeping up the mountain sides, and to stand, rod in hand, in some lonely glen, gazing at the beauty of the scene, as bursts of glowing sunshine are thrown from Nature's lantern upon sheets of mist and mellow-tinted mountain sides, to view the jutting crag o'er which the raven croaks her echoed call of warning, and to see the gaunt pine trees stand forth amid the whirling mass; to see them growing clearer, more defined, as Sol, more powerful, drives his weaker foe into oblivion ; ah ! that foe is "cornered" here, and in his power, and before him must melt away; not so in the great city, where the mist is made by human beings into something else, and even Sol in all his grandeur often cannot lift the curtain when it falls upon St. Paul's and Westminster. But here, to stand and see it lifting, leaving behind a view on which 'tis good to gaze, the moor, the rocks, the trees, the mountains hanging over all, and down below a peaceful valley with its river winding far till lost in the expanse of distance.

But at our feet a brook, its mossy sides and rocky buttresses reflected in the still clear pools, the gossamer upon its banks still hung with dew-drops, and the plants upon its margin with their heads still hanging down as if in slumber. Sol has not touched them with his finger tips as yet. To gaze on such a scene inspires the town-born traveller with ecstacy, and a feeling as of awe and wonder rises within him, mixed with keen delight, as the water of the brook beneath him circles round a given point where *Salmo* touched its surface. A little one, 'tis true, but t'was a *Salmo* really. See! there another rises.

The traveller smiles a pleasant smile, brings forth his tackle, and essays to tempt the little fish into his creel—at the third cast he is successful, and steps lightly on to the next pool to try again —another fingerling is thrown upon the bank, and yet another, lured from its watery home. And this the traveller is content to call his fishing, and to view all things round as exquisitely beautiful; more beautiful because he holds a fishing-rod—it helps him to enjoy the scene most thoroughly and to make the best of everything.

Then how much more, when climbing yonder bank he comes upon a pool larger than all the rest, and deeper too, a pool which human hands have made, and stocked; and here he tries his "gentle art"—soon has a rise—is into one, "A monster" quoth our tyro. Off goes the fish, out runs the line till thirty yards are gone, then slackens, and he reels him in awhile. But off he goes again, now plunges and then leaps from out the water, shewing for a moment his bright silvery scales—a full three-pounder surely—ah! yes, he turns the scale at that when he is brought to bank at last, and safely landed. Another cast or two and then another fish is hooked, more playful even than the last; he plays, he fights, leaps, rushes, and lies panting in the net, then in the creel.

But why the change from fingerlings to what you may call fish ?

It is because that wonderful provision long ago designed by Nature for mankind, giving him dominion over all her creatures, whether beast, or fish, or fowl, has been made use of, by which the water of a simple moorland rill can now with human aid produce enormously.

It must be quite apparent, even to the most casual observer of the laws of Nature, that there exists a wonderful balance of animal and vegetable life, which has been kept up for ages, by the destruction of one species by another.   Man has the power given to him of altering that balance, and of adjusting Nature's laws to meet his own requirements.   Interference with Nature's balance, however, is a matter which should receive serious consideration before action is commenced.   Experience teaches us undoubtedly that where man thoughtlessly interferes with it the result means loss to himself, and as an example of this we may take the rabbit pest at the Antipodes.

Where thoughtfully done, however, the result is often one of great benefit.   Of this there are a good many examples, both in the animal and vegetable kingdoms.   The axiom we know applies very largely to the cultivation of the land, and it also applies in an even greater ratio to the cultivation of the water.   Man has now the means of dealing with it in such a manner, that it is quite practicable to utilize the small rills of our mountain or lowland valleys, and make them produce an abundance of large fish, where the merest dwarfs existed before.*   It has been done most successfully, and may be done in thousands of other places in this country, and often at a comparatively trifling cost.

*In the case of many streams there is, unfortunately, a very bad reason for the fish never or only rarely exceeding a certain size ; and that is that nearly every one above, say, a quarter of a pound in weight is taken out by the poacher. He fishes chiefly with a net which will not take the smaller fish, but which is most destructive amongst large ones. Often have I come across him, or his shadow, out at night, but as long as the law winks at such proceedings, by imposing penalties, when caught, at which he chuckles, so long will he continue to depopulate our streams. The net used in this part is usually the shackle or bag-net. I have sometimes used one for obtaining spawners from a stream,

A great deal has been said and written about the water area of the globe, but when we take into consideration the water facility which we posses in Britain, and realize that it is a great motive power which may be turned to excellent account, the wonder will be that it has been so long allowed to run idly down our hill sides, and along our valleys, where it might have been made, by means of a little ingenuity and engineering skill, to do good work in far more ways than one. I have briefly described a portion of a natural mountain or moorland stream, and nearly everyone who makes any pretence at trout fishing, is acquainted with such a stream in some shape or other. It may be a dashing mountain torrent, careering along over its rocky and boulder bestrewn bed through wastes of heather; or perhaps at a lower elevation, it may be passing more sluggishly between earthy banks cut through a fertile tract of country; but it is still the same in one respect—it is a trout stream. Everyone who has been accustomed to frequent the banks of such a stream needs not to be reminded how, in a great many instances, trout of any size are rare. The stream may be replenished by the introduction of other and better breeds, and although this undoubtedly does good, it is necessary that other things be equal, or the result often discourages rather than otherwise, a matter on which I shall have more to say in my next chapter. But by dealing with the motive power which Nature has provided, and making a series of artificial pools which may be kept under absolute control, the stream containing only

and can testify to the destruction that may be wrought with such a deadly engine when in improper hands. In private or enclosed waters the matter is now fortunately very different, and several good stiff penalties have been imposed. The case then becomes one of stealing, and will carry a very much higher penalty than that inflicted for poaching. In many cases which come under the latter head, absolute encouragement of the crime has resulted from the absurdly inadequate fines which have been imposed. I have known cases of men who have made a living by poaching, and who, when caught once in a while, have been fined a few shillings, and have within a few hours not only made up the fine, but a good deal more, by the plunder which they have obtained on another robbing expedition. In the interest of the men themselves, the sooner such practices are put a stop to the better. In a case which was brought before the Sheriff at Forfar of a man stealing fish from a private pond, evidence was given shewing that a certain proprietor had stocked the pond with trout. In such a case the trout, having been put into an enclosed piece of water, are as much the property of the owner as a flock of sheep, and this should be made widely known. The Sheriff said that the case involved a curious point in law. Under some circumstances, a person found fishing for trout could not be charged with theft; but amid the circumstances connected with the particular case, anyone who caught fish in the pond without the proprietor's sanction committed theft in the same way as if he stole the trout from a bowl.

fingerlings may be made to produce big fish that are worth the catching, worth a journey, and without which we cannot make " An Angler's Paradise "

CHAPTER III

Frank Buckland—His prophecies—Their fulfilment—Troutdale Fishery— Introduction of black bass and American trout—Solway Fishery commenced—Its progress—Vocturnal adventures—Discovery in Germany by Golstein—Jacobs— Gehin and Remy—M Coste—Huningen—Gremas—German Progress

LITERATURE tells us of the fish culture of the olden times, the esteem in which fish were held by the ancient Greeks and Romans, the ponds of the monks in Great Britain, and how the Chinese ingeniously collect the spawn of fishes on bundles of sticks and mats placed in the water, and how it is sold in their markets   On some parts of the Continent, too, fish are taken alive to market, and those which are not sold are taken back and returned to their pond living and well

As a pioneer of fish culture in this country there is no more honoured name than that of the late lamented Frank Buckland How well I remember, some thirty years ago, listening to his talk about trout and salmon, and their ova, and reading his book on " Fish Hatching," published in 1863   More than a quarter of a century has rolled away since those days, and it is exceedingly interesting to look back and see how largely the work has developed since that time, often progressing under considerable difficulties, until it has reached its present magnitude

Frank Buckland said of fish culture that it promised "to be eventually the origin or increase of revenue to private individuals, a source of national wealth, and certainly a great boon to the public in general "

The first portion of the prophecy has been fulfilled , the second is only waiting to be so, as soon as our Government will step in and take the matter in hand with regard to some of our marine and anadromous fishes, or else by suitable laws make the way easy for private individuals to do so. The third part of the prophecy is now being fulfilled, and the time will soon be when it will be said that fish culture is "a great boon to the public in general."

It was this meeting with Buckland, coupled with a great love for Nature, and a strong desire to make its study of some practical use to myself and to my fellow-men, that first set me to work hatching fish ova. The first experiments were tried in a small apparatus rigged up over the water tank in my father's conservatory, and which resulted in trout being grown to a quarter of a pound in a cellar close by, and the subsequent erection of a small hatchery in his grounds, where trout ova were successfully dealt with, and the fish reared for several years. Finding this place and its water supply too small, a site was finally selected, in the year 1868, among the Cumberland mountains, for the first real hatchery ever erected in this country on commercial principles. The work at this, the Troutdale Hatchery was, owing to the nature of the surroundings, only carried on upon what would be considered now a very limited scale, and for twelve years under considerable difficulties, my time being closely occupied more than a hundred miles away, and it was only an occasional visit that I could give to the fishery and its work. In 1880, I was liberated in an unexpected manner, to carry on and devote my whole energy to the work, and I have now the satisfaction of looking upon a most successful issue to my labours.

In the working of the establishment in Borrowdale, among the Cumberland mountains, I was assisted by the late John Parnaby, of Rothwell Haigh, in Yorkshire, who had just returned from Canada, where he had for some years been engaged in fish-cultural work under the Canadian Government. His experience was considerable, and coupled with my own knowledge of the subject, and a love of the work, we soon had a good stock of fish. Parnaby made several voyages to America, for the purpose of increasing his practical knowledge of fish culture, and of bringing to this country some of the more valuable food fishes of that continent. The first living black bass *(Grystes nigricans)* ever seen in Britain, were brought over by him in 1873, when I met him on board the ship in Liverpool, and helped to convey the little fellows to the Troutdale ponds. For this work we were rewarded by receiving the silver medal of the Société d' Acclimatation de Paris.

A number of these fish, weighing about a pound each, were safely brought by Parnaby the year before (1872), as far as the Irish Coast, where they were simply battered to death in a terrific gale which was encountered off the Fastnet Rock. They were landed in Liverpool the next day perfectly fresh, and two of them we ate, and gave the rest away, one being sent to Frank Buckland for his valuable collection. The two which Parnaby and I disposed of proved excellent.

In the year 1869 we introduced the American trout *(Salmo fontinalis)* into this country, and soon had a fine stock of these fish, which did exceedingly well in the ponds at Troutdale. Since those days they have been distributed through the country, and in some waters have done remarkably well, whereas in others they seem to have disappeared. The migratory instinct in these fish is very strong, and at certain times of the year they will leave a lake or pond and push up stream, or down, as the case may be. They go to the sea, and have been caught in the salt water, in some of our bays and estuaries. This sea-going habit alone proves them to be good fish, but it renders special precautions necessary in order to prevent them from making their exit. Where such steps are taken, by the simple fixing of a screen at the outlet of a pond or lake, the fish are easily retained, and in many instances have given great satisfaction to their owners.

In other cases, where they have had free access to a river, they have simply run away. That the *S. fontinalis* is a real game fish is beyond question, but that it is not adapted for all our waters seems to be also a settled fact. In some waters it is accused of not rising to the fly, but I have not yet met with such a case personally. I have on the other hand made its acquaintance in many places, where it rises in a manner that has astonished many old and skilled fishermen. Where these fish can be kept in a lake, and allowed access to a set of artificial spawning beds, which will be described in my chapter on the construction of fish ponds, they will do well. For a pond near a house they do admirably, and are excellent eating when well fed in suitable waters. No river in this country has ever yet been stocked with them. The turning in of a few thousand fry is not stocking a river. Until 250,000 eyed ova have been planted for three consecutive seasons, we have no right to pronounce the stocking of British rivers a failure. Two years ago, I offered to bear half the cost of the experiment on certain conditions, in the hope that some large proprietor would come forward and join in it, but the offer was not accepted. Until something of this kind has been done, it is premature to condemn the fish as useless for our rivers.

The work at the Troutdale Fishery was continued until John Parnaby's death, when I took the whole burden of it on to my own shoulders, and finding the available space and the water supply wholly inadequate, I began searching for another site. No one could imagine the difficulty experienced in finding a really suitable place for the erection of a hatchery, without having gone through it. However, the right spot was found at last, and upon it now stands the well-known Solway Fishery.

Having secured the necessary land, the first step taken was to dig out five ponds and erect a hatchery. The latter was built of granite, eighty feet long by twenty feet wide inside, and has since been much enlarged. Hatching boxes were fitted up in this room, and one corner was partitioned off for an office *pro. tem.* At this time I was living fifteen miles away, and found that as the work grew my presence was needed on the spot, and therefore I took up my quarters at the hatchery. It was situated in a wild remote corner among the moors, with only one little cottage house in sight, and there being no sort of accommodation whatever, I did as I should have done had I been in America—I camped.

Yes! a whole winter was spent in that hatchery, and a very enjoyable winter it was. Before the next I had built a shanty, in which to live when my presence was required on the spot, and this has since been added to until it has assumed considerable proportions. Whilst engaged in the development of the fishery, there were, as might have been expected, sundry little bits of adventure, which rather "added a spice to the cake" than otherwise. Perhaps the most remarkable was an occurrence that took place on the night of December 11th, 1883, when a terrific gale, of greater force than I had before experienced, burst upon us.

Soon after midnight the storm began to reach its height, and at one o'clock in the morning the thick plate-glass of a large window was blown in, the frame being left intact—a few pieces of broken glass still remaining in it, while the rest were scattered over the room, a considerable quantity being found in the fire-place. Hastily calling an assistant, who was soon on the spot, we set to work to block up the broken window, in order to prevent the wind from getting into the house and doing further damage.

I remembered a wooden platform that had recently come back from the International Fisheries Exhibition at South Kensington, and we ran to the place where it was lying along with some large cases of hatching apparatus, models, etc., still unpacked, but were met by one of the cases which came careering through the air, passing us within a few feet. The other cases had already gone, and the platform would soon have followed had we not secured it. As it was we had not much difficulty in getting it carried to the window, for we had almost a fair wind, and by steering a little we kept a good course, the only thing needful being to let go just at the right point. To have taken it back would have been impossible for six men. It was speedily fastened up and blocked the aperture safely.

The wind at this time (about 1-30 a.m.) was terrific. I have been in gales both on sea and land, but never witnessed anything approaching this. Every now and again there was a complete lull, and we could hear the wind sweeping down the pine woods in the distance, with a roar as of a mighty flood, and as it struck the house the noise was as if some huge battering ram had been brought to bear upon it. Large timber trees were uprooted wholesale, or snapped asunder, and daylight revealed the awful destruction that had been going on in the woods around, many thousands of large timber trees being blown down. Many of the woods were lying quite flat, the trees torn up by the roots, as if some giant hand had been doing a little weeding, and after pulling them up, had left them to wither and die.

Fortunately the night was clear and bright moonlight, not a cloud being visible, and we soon found the relics of the exhibition; the wind had carried them until they stuck in some trees and dropped into a brook, which, being full, carried them off, but fortunately they stranded or stuck fast in bushes before travelling very far. We could do little towards getting them out then, so allowed them to remain till morning, when they were duly recovered. I felt rather uneasy about the spouting that conveyed the water to the hatchery, and we went round to the back of the building to inspect it.

The spouts had been firmly nailed to stout oak tressels fixed in the ground. The tressels were immovable, but the spouts had parted company, and were simply "to seek," to use a common Yorkshire expression. It seemed to come in very appropriately here, for we literally had "to seek," and finally found the spouts, sticking in some trees near, well to leeward of course. We soon had them out and commenced to carry them back, but a gust of wind took them out of our hands and overhead back again from where we had brought them. After procuring a hammer and some stout nails, we again commenced carrying the spouts one at a time, and by dodging the wind, got one of them on to its tressels and got in four nails. While we were carrying the next length of spouting, however, we saw the first, which we thought we had firmly fixed, flung off its seat by the wind, and the second was no sooner fixed than it was served in the same manner.

Just at this moment came a terrific gust which lifted me off my feet, and but for taking a regular dive into the wind as if it had been so much water, I should have been carried into the stream close by, or possibly into the trees. At the same time the crashing of timber in the wood above us was terrific, as upwards of two hundred splendid larches fell flat as the walls of Jericho, knocking each other down like ninepins. The air was filled with flying branches, sticks, and other missiles, and suddenly a cloud rose from the earth a short way off, and obscured the sky for some distance, finally losing itself in a plantation of young Scotch firs.

The effect was remarkable, and as another gust came down upon us, a second cloud proceeded on the same course as the first. Examination proved that a haystack had migrated, and the appearance of the fir plantation the next morning was as if all the sparrows in the three kingdoms had built their nests in it, while of the haystack only some two and a half feet of the bottom part remained. Finding we were absolutely powerless to cope with such a storm, and feeling sure it could not last long, we went indoors, and by this time being fairly hungry, we soon had a good meal prepared, after partaking of which we drew to the fire and smoked the pipe of peace, while the storm continued to howl on, the noise being at times almost deafening.

The conversation naturally turned on storms, and several memorable ones were discussed, but none of them would bear comparison with the one that raged that night. I thoroughly enjoyed it, but at the same time hope I shall never see such another. By six a.m. the wind had moderated sufficiently for us to get the spouting fixed, and the water was again turned on to the hatchery. It had been cut off for five hours, and I was rather afraid some of the ova would have suffered, but everything went well, and I never could trace any ill effect to this memorable night. In the morning several of the natives kindly came some distance to render assistance, expecting to find the "shanty" a heap of ruins, and I had a good laugh as one of them seriously told me about this some time afterwards.

The water is now, and has for some time been conveyed in underground pipes, and the arrangements are most complete; no wind or frost having any influence on the regular flow into the hatcheries. For many years all the spawning was done out of doors, but now in bad weather a good deal is done in a spawning house. This is conducive not only to the comfort of the workers, but to the welfare of the ova, as a better impregnation can be got by keeping wind and rain and sleet out of the spawning dish, and everything as dry as possible. In addition to this the days are very short in Scotland at spawning time, and with a spawning house the work can go on all night when necessary, without the slightest difficulty.

The discovery in modern times in Europe, of the art of

hatching fish ova by artificial means originated in Germany, and is ascribed to a German naturalist, Count von Golstein, who is said to have made the discovery in the year 1758. Having obtained some ova and taken what he considered to be the necessary steps for its preservation, he was rewarded, when the proper time had elapsed, by seeing young fish produced. Later still, about the year 1763, one Jacobi, another German naturalist, performed the same experiment, with this difference, that he took his eggs from a dead fish instead of a living one; and what is more remarkable, his experiment also proved successful.

Bertram, in his "Harvest of the Sea," says :—"Jacobi, who practised the art for thirty years, was not satisfied with the mere discovery, but at once turned what he had discovered to practical account, and in the time of Jacobi great attention was devoted to pisciculture by various gentlemen of scientific eminence. . . The results arrived at by Jacobi were of vast importance, and obtained not only the recognition of his Government, but also the more solid reward of a pension." The well known Spallanzani also experimented in Italy on the ova of fish, as well as upon the spawn of toads; and since then many other experiments have been tried by scientists and others in various countries, but no one ever seemed to think of turning the knowledge gained to any practical account, but merely looked on the whole affair as an interesting scientific experiment.

Although discovered in Germany by Count von Golstein and by Jacobi, the art of pisciculture became known to two humble peasants named Gehin and Remy, in France, about the year 1840. These two men lived in an obscure village called La Bresse, in the department of the Vosges, where it was observed that the supply of fish was falling off. They found out how the eggs of trout were fecundated, and following up the knowledge they had gained, in 1841 they succeeded in hatching their first eggs in a very rude sort of apparatus placed in the bed of a stream. During the next three seasons they continued their work, and in 1844 were presented with a bronze medal and a sum of money by the Société d' Emulation des Vosges, as some encouragement to their praiseworthy efforts.

These two poor men knew nothing of Count von Golstein,

or of Jacobi, having probably never heard of either of them, nor even read a book on Natural History in their lives. Nothing was heard of the discovery beyond the department of the Vosges, before the year 1849, when Dr. Haxo, of Epinal, secretary to the Société d' Emulation, and member of the Conseil Académique of the Department of the Vosges, sent a communication to the Academy of Sciences at Paris, describing the method of cultivating fish, and it sorely puzzled many who heard it, that it should have fallen to the lot of the two poor fishermen to put into practice and show the value of a discovery which had been known to many learned men for a long series of years.

The subject was at once warmly taken up by the Academy, who, seeing the great importance of the discovery, lost no time in calling the attention of Government to it. The Government decided to carry on the process on many of the rivers of France, and Gehin and Remy were sent for and employed at good salaries; and a Commission, consisting of a number of scientific men, was also appointed to superintend their operations. The plan adopted by these two men for the artificial cultivation of trout, was to procure a number of round boxes made of zinc, in shape somewhat like a cheese, and about eight inches in diameter. These were riddled with small holes, and in each of the boxes was placed a layer of fine gravel, and upon this the eggs were laid. The boxes were then placed in the bed of a stream and covered with loose pebbles, and the water was allowed to percolate through them by means of the small holes. The young fish on being hatched were kept in these receptacles from eight to fifteen days, and then set at liberty.

Although it was long considered that the gravelly bed of a natural stream was a requisite in fish hatching, which could not be substituted, an eminent French naturalist, M. Coste, Professor of the College de France, at Paris, discovered that it might to a certain extent be done without, and he proved his assertion by producing salmon in a tub. Having procured a large tub he had a number of small conduits or canals constructed in it, and placed in such a position that the water flowed from one to the other, and at last, when its services were no longer required, escaped from the vessel altogether. In each of these vessels he placed a layer of gravel, and laid upon it a quantity of salmon eggs. His supply of water came from a cistern, and flowed through an ordinary tap, the only precaution being to keep the stream constantly going, and this, it may as well be stated, is an absolute necessity, if trout or salmon ova are to be successfully hatched.

In the year 1854 Professor Coste founded the important fish-cultural establishment so well known as Huningen, and by its means large quantities of fish ova were distributed. Being situated in Alsace it fell into the hands of the German Government in 1871, and under the directorship of Mr. Herman Haack has since then, as before, done very good work. The hatchery and ponds are said to cover upwards of eighty acres, which gives some idea of the extent of the work. A large space in the chief building is devoted to the hatching of the ova of the *Salmonidæ*, but ova of many other kinds are dealt with. From half-a-million to a million salmon ova are hatched each season for stocking the waters of the Rhine, which is near, and the Government pays for this work.

The young salmon are reared in concrete tanks, which are also used for other fish, and answer the purpose admirably. The Governments of Germany, Switzerland, and the Netherlands, have shewn their interest in the important work of fish-culture, which has been going on for some years, and it is largely due to this fact that it has made such rapid strides in these countries. Public waters are being restocked and private enterprise is encouraged. In Saxony fish-culture has been helped by courses of lectures which have been delivered by Professor Nitsche of the Academy of Forestry. The lectures, illustrated by diagrams, specimens and apparatus, have dealt with the subject in a very exhaustive manner.

The knowledge which had been gained was by degrees increased and fish-culture has prospered in France, until to-day there exists at Gremaz in the Department of the Ain, one of the most perfect systems of fish-culture of the present time. Fish are reared on the natural food produced in the water, and herein lies one of the great secrets of success in the rearing of trout fry.

In Germany, too, fish culture is being carried on with considerable success, and trout are now extensively grown for the market. The late Herr Max von dem Borne did much for fish culture in Germany. On his estate of Berneuchen in the Province of Brandenburg, he commenced a fish-cultural establishment in 1876, a short description of which may be interesting. The water supply is drawn from a stream known as the Mietzel by means of an aqueduct, and it is thus exposed to considerable variations of temperature. During winter it goes down to freezing point, whereas in summer it goes up to 70° F., and occasionally a temperature as high as 77° F. is reached.

It is carefully filtered before being allowed to do duty in the hatching tanks, which are made of concrete, and are used for hatching the ova of *Salmonidæ*. These tanks are provided with covers. Eggs of the *Coregonidæ* are also dealt with in a special apparatus designed by the proprietor. Six different American *Salmonidæ* have been introduced, as well as the black bass (*Grystes nigricans*), the American catfish (*Amiurus nebulosus*), and others. There are now many other establishments prospering in the country. That of Herr Siegfried Jaffé at Sandfort is a thriving place with thirty ponds or more, and an extensive hatchery where many kinds of trout are successfully cultivated. Mr. Jaffé tells me that a very large number of trout are sent to market in a year for eating purposes, and they are only grown up to a comparatively small size, as they are then more valuable as articles of food, being younger and more tender, and possessing a finer flavour. With regard to the eating quality of trout, for which purpose a large number are killed annually at the Solway Fishery, I can quite bear this out, the best flavoured trout being those running from half a pound to a pound.

Strange it is that man, who has tunnelled through mountains, who has bored into the depths of the earth in search of mineral treasures, who has drained many a treacherous swamp, and caused waste and barren lands to yield plenteous crops of corn, or turned them into rich pastures; who has invented steam engines, and by their means been enabled to traverse oceans and continents in an incredibly short space of time; who has even made the tremendous agent electricity itself subservient to his will, and made so many

D

wonderful mechanical contrivances, should have almost entirely neglected, until within the last few years, to endeavour to increase the rapidly diminishing and deteriorating supplies of fish in some of our waters.

True it is, that as the mournful looking morass or the heathery moor have in so many cases been transformed into fields of waving corn, and often made to produce an abundant crop where hardly a blade would grow before, so our most unfruitful streams can be filled with salmon or trout; our lakes, ponds, reservoirs and other waters may be made to teem with life, and to produce as rich and abundant a harvest as some of the best cultivated fields in our favoured land. And why not cultivate fish just as much as ducks or poultry, or any other produce of the farm-yard?

Now that we know that a river or a pond may be doubled in value by a reasonable outlay; or, to put the matter in another way, now that we know that the yield of fish may thus be doubled or more than doubled in any locality, surely the matter is worth our serious attention. Whether entered into with a view to supplying the market, or with the intention of affording an increase of sport to anglers, the time has come when the growing of fish no longer remains as a curious and interesting scientific experiment, but is *un fait accompli.* Much has already been learned, as the results are proving, and there is great encouragement in looking back for twenty years and noting the progress that has been made. There is a wide field open before us. The only question is—Who will go in and occupy it? There is endless scope for development of our water, and now that we know how to develop there is good reason for making a commencement. Every valley should have its set of fish ponds for angling purposes, and be able to provide some good fishing for those who want it.

The fish ponds that are at present to be found on various parts of the Continent, are not only highly interesting, but are also very suggestive. As far back as the middle ages, the monks are said to have had a valuable system of fish culture. In this country, too, we know that it was the case, for not only is it an oft-repeated historial fact, but the remains of the fish ponds are there to tell their own tale. What was the special object of the monks in possessing such ponds, and in cultivating fish, when there were plenty in the rivers and other waters not far off, as there were in those days? It was that they might have a certain supply instead of an uncertain one, and that they might have them just where they wanted them and as they wanted them, and they had them. And so may we have our fish ponds well stocked with magnificent trout as we want them, where we want them, and when we want them.

CHAPTER IV.

Referring to Lake Vyrnwy—Loch Leven—The English Lake District.

IN the merry month of May, 1891, I visited the well-known Lake Vyrnwy, in North Wales. It was at that time a piece of virgin water, so to speak, and I had a good opportunity of seeing it under somewhat varied circumstances as regards weather, etc.,—so important a factor in any work one has to do where trout are concerned.

The lake was made by order of the people of Liverpool, primarily for the purpose of obtaining an increased supply of water for drinking and other purposes. I fear I rather shocked a respected inhabitant of that wealthy city, who was evidently not an angler, and who expounded to me some of the advantages possessed by their already celebrated new waterworks, and then turning to me, asked for my opinion. "Why!" I said, "you've made the finest fish pond in the world!"

And so it is, or may be, and if the reader will have a little patience, I will endeavour to describe this charming piece of water, and some of its surroundings.

It is to be found in the parish of Llanwddyn, in Montgomeryshire, and is some twelve miles distant from the nearest railway station, Llanfyllin. The exact distance as laid out on the ordnance map is something less, but anyone who has driven over the road will, I think, agree that the local reckoning is not out of place, taking into consideration the elevations reached in traversing the wild hills that lie between the railway terminus of Llanfyllin and the magnificent sheet of water now known as Vyrnwy Lake. On the occasion of my visit, I travelled from Scotland through the night, and reached my railway destination, *via* Crewe and Llanymynech, about 9-20 a.m. I may say that I travelled in a special fish car conveying over twenty thousand trout for the lake, being part of a much larger number put into it by the energetic lessee, G. Ward, Esq., of Bala, whose genial face was visible on the platform as the train glided alongside with its precious freight.

I had breakfasted *en route*, and therefore, after seeing the fish carefully transferred to the waggons that were waiting for them, Mr. Ward and I drove off, to do the twelve miles that intervened between us and our destination. The air was fresh and balmy and still smelt of the morning, as it was wafted, by the delightfully cooling breezes, round hilly corners covered with enchanting foliage, and over flower-decked meads hemmed in by densely wooded slopes. Between the little chats on fish and kindred matters, which which we enlivened the journey, I had time to look around and to enjoy the beauty and freshness of the scene. The plaintive cooing of the dove *(Columba palumbus)* was heard as we passed through belts of woodland, and at one bend of the road the harsh scream of the jay *(Garrulus glandarius)* resounded through groves of oak and hazel, while the little tits *(Parus major* and *cæruleus)* cried "ze zrr," as they hung suspended from the twigs on which they searched the buds for insects. All Nature seemed alive and in her gayest garb. Even the butterflies upon the flowers looked fresh and beautiful, except *Vanessa*, fluttering round a stone heap ; where she flickered her bedraggled wings, that told of hybernation and the clammy chills of winter, past and gone. Beneath us lay a trout stream, in places almost hidden from the gaze of man by vegetation of the richest kind that grows in those parts.

Our conversation was of trout, varied with little bits of history, natural and otherwise, but like a famous member of the feline race, of which I've heard in song, it would "come back" again to trout, their culture and their means of capture. Thus we found ourselves at length upon the mountains, where the curlew called at distant intervals. We crossed the dashing moorland stream, on which the dipper makes his home. The height we have come over is upwards of 1,000 feet above the level of the sea, and the descent into the valley, which we now must cross, is sharp, and by a zig-zag path cut out of a hill side. 'Tis well our leader is sure footed, for as it is, the traveller who is unaccustomed to such roads, is wont to hold on fast to something; but there is no need for this. To him I say, sit still, and view the grandeur of the scene, enjoy thyself, breathe freely, and reflect upon the treat in store for thee, on Vyrnwy's placid bosom, 'ere the sun goes down.

Another range of hills is covered, and we are down into the next valley, from which, however, we emerge at length. Before we quit this pleasant vale, in which we ourselves are hidden, and where the neighbouring mountains are also kept out of view by the closeness of the wooded hills that bound it, we get a sight of the embankment, but passing on we turn sharply to the right, and winding up a steep ascent, presently find ourselves on level ground again.

Before us bursts suddenly a view, the grandeur of which we may travel far to surpass. It comes upon us as if by some magic process, like the lifting of an enormous curtain. In reality we emerge from a sort of semi-natural shrubbery, and stand upon a gravelled plateau. Beneath us lies the lake that we have come to see; to the north the dark and sombre looking Berwyn mountains; above, blue sky and fleecy clouds, which cast their flitting shadows on the lake and hills around. Save where these shadows flew along, the whole was bathed in sunshine. It also shone upon a dark grey pile of structure, made by human hands upon the lake itself, looking like some grim watch tower standing there to tell its tale of bye-gone days; but no, it is the Vyrnwy Tower, standing like a castle on the Rhine, otherwise the straining tower.

It stands in fifty feet of water, and is reached by a causeway on four arches, and in height is upwards of a hundred feet above the water. It is used for straining the latter, a most important part of the preparation of the liquid for the natural wants of man, whose requirements demand that it should be as pure and free from sediment, as if for hatching trout ova. The process that is adopted is indeed good for both purposes, and is comparatively simple: the water being passed through screens, covered with fine copper wire gauze, of over 14,000 meshes to the square inch. These screens are removed and cleaned, as they become clogged with the matter held in suspension in the water, and in this, very much resemble, in their mode of working, the screens used in the filtering apparatus of some fish hatcheries.

After gazing for a few moments at the grand display of scenic nature, we hurried in to luncheon, to which we both did ample justice. The lake may be approached from Bala if required, the drive being fifteen miles to the hotel, which is replete with every comfort, and I cannot speak too highly of the reception which we had. Truly, it is an "Angler's Paradise." The spread of shining trout upon the table in the hall was a sight worth going far to see. Each basket was weighed whilst the guests were at dinner, and after a sumptuous repast. we strolled to witness the display. The examination of the fish, the discussions which arose, the friendly chat that followed, such as is only known to anglers, all combined to make a pleasant evening, after which each guest retired to rest, and pleasant dreams of giant trout drawn from the hidden depths of Vyrnwy.

The lake was made by building an enormous dam or bank, which is in reality one gigantic block of concrete. The mode of its construction is interesting. The stone used was obtained from a quarry near, and is hard and dark in colour. It belongs to the Caradoc group of the lower Silurian system. Some of the blocks used weighed as much as ten tons, and above a third of the stones of which this huge dam is formed weighed over four tons apiece. Each stone was worked to a flat surface on the underside, and was then lowered into a bed of concrete, and the inter spaces filled with concrete, which was well rammed, in order the better to consolidate the mass.

The dimensions of the lake and dam are as follows :—

Height from bottom of foundation to the parapet of roadway	161 feet.
Length along the top	1,165 feet.
Height from river bed to sill of overflow	84 feet.
Greatest thickness at base	120 feet.

Batter or slope—

Inside the dam is	1 horizontal in 7¼ vertical.
Outside ,, ...	1 ,, ,, 1½ vertical.
Superficial area of the lake	1,121 acres.
Average width	½ a mile.
Circumference	11¾ miles.

There is a road over the dam, which is supported by a viaduct, consisting of thirty-one arches, each of twenty-four feet span. · Under these arches flows the surplus water when the dam is full, and at such times it makes a miniature Niagara, and adds a charm to the valley. The bottom of the outer face of the dam is built in a curve, so that the water strikes it at an angle, and does no harm.

One of the most important features of the dam, from an angler's point of view, is the means adopted by which daily compensation water is provided for the benefit of the river below. This water is discharged by a pipe eighteen inches in diameter, and there is another pipe thirty inches in diameter for discharging monthly compensation water. This is the water which the Liverpool Corporation has to turn into the river Verniew, for the satisfaction of riparian owners and the Severn Navigation Board, as compensation for impounding the head waters of the river. More than this : the Corporation also has to deliver to the satisfaction of the Severn Fishery Board, twelve hundred million gallons of water annually, whenever it may be desired, in the form of freshets, each freshet to consist of forty million gallons.

This is a condition that is imposed by Act of Parliament, and must be a great boon to the inhabitants of the valley. Before this reservoir was constructed, floods used to sweep down the vale, carrying away crops, mill weirs, etc., and doing sometimes immense damage to the land and property owners and their tenants, whereas now a very large proportion of this flood-water is retained, and allowed to flow off gradually. The great advantage of this is apparent, as well as compensation water during dry weather. There is one very important point which has been overlooked in making the arrangements for the supply of this compensation water. It is drawn from far too low a level. Had a competent fish culturist been consulted, such an arrangement would never for a moment have been considered. The water drawn from such a point is known to be so far unsuitable, that such an oversight would have been guarded against, and proper measures taken for drawing it from the right point.

Many of our streams could be dealt with in this way with very great advantage. A reservoir, or a series of them, in many situations, would be a good investment if stocked with trout, whilst the compensation water, in times of drought, would be an untold advantage to the stream below; to say nothing of the advantage to the miller, or the ratepayers of a town upon its banks. The angler would certainly have cause for rejoicing, and the increased riparian value would be considerable. The question, too, of food supply for the trout, would then be much more easily dealt with, a matter of the most vital importance in dealing with our moorland and other streams.

The Vyrnwy lake itself is four and three quarter miles in length, and is fed by several streams, which are available for, and provide good natural spawning ground for the trout, a matter of no small importance. The spawning ground, too, is near the lake, and the fish are prevented from running too far up these streams, so that during the spawning season they can be closely watched, which is another great advantage. On one of these tributaries, or feeders, a set of rearing ponds has been made, and fish are hatched and grown in them, for keeping up the supply in the lake. The natural food of the fish is also cared for, as well as the aquatic vegetation, so absolutely an essential in a lake like this. The depth of water on the embankment is eighty-four feet, and for a considerable distance up the lake there is very little variation, the bottom of the valley having been nearly level.

There seems to be little doubt that this valley was in former times occupied by an ancient lake, and that the *débris* brought down from the surrounding mountains, in course of time, filled it up. The deposits of successive floods helped to form a fertile crust of alluvial soil, and at the time the dam was made, a good part of the valley was under cultivation. In the course of filling, the vegetation is submerged, and a large portion of this goes on existing for some time; nay, in places, even goes on growing, whilst at the same time an enormous amount of decay naturally sets in. By means of this decay, favourable conditions are produced for the multiplication, to an enormous extent, of a large number of creatures on which trout feed and thrive.

This accounts for virgin waters giving excellent results, as regards their fishing; whilst in a few years there is a great falling off, both in the quality of the fish and in their size, unless steps are taken in the meantime to prevent it. Even though this be done, it may not altogether counteract the reaction which takes place

when the submerged vegetation disappears, and before its place is
supplied by the aquatic forms so necessary in a lake where a good
stock of trout is to be maintained. All this was carefully considered at Vyrnwy, but notwithstanding the care which has been
taken to retain the necessary favourable conditions, the year 1892
saw a decided falling off in the fishing. It is only fair, however,
to the lake, to say that during 1893 the fishing improved again,
and the catch was greater than ever it had been before. It is,
therefore, quite likely that Vyrnwy may "beat the record" in this
respect, for I have hardly known an instance of a newly made
lake that did not shew a falling off after the first results, which are
usually good.

The summer of 1893 was unusually dry, and the result was
that the water level lowered, until in places a considerable amount
of foreshore was left bare. Upon parts of this a good crop of
vegetation sprung up, and on the rise of the water this was
submerged, and it is a noteworthy fact that some of the best
fishing ground is in the neighbourhood of this submerged
vegetable growth. A very important lesson is to be learned from
this, for it teaches us clearly, that even the lowering of the water
in a reservoir, by which the banks are left dry, may be turned to
good account, by the sowing of a suitable crop of vegetation.

It certainly may be made highly beneficial, by building a few
retaining banks, behind which water would remain, and which
might be made exceedingly valuable as food producing grounds,
and which would, to a large extent, tend to compensate for the
laying bare of the foreshore.

The well known fishery of Loch Leven is another example
of what might be done by cultivating a piece of water. Prior to
1830, this loch is said to have been upwards of 4,300 acres in
extent; but during that year, a drainage scheme which was carried
out reduced it to 3,543 acres, lowering the level of the water
permanently some four and a half feet, and leaving a barren
margin. A few years later, the fishing of the loch was fouud to
have been seriously injured by these operations, and it was
calculated that it had been reduced quite one third in value. A
more natural result could not have been, and a large number of
our lakes have suffered in the same way, the tendency having

been in the past to drain away the waters and lower the levels of lakes in many localities.

In a great many instances where similar arrangements have been carried out, considerable damage to the fisheries has resulted. The doubtful quality of the land reclaimed, in many cases, has not at all compensated for the cost of the work, plus the amount of depreciation which has accrued to the fishing. The levels of some of these lakes are now being raised again, and some of the land adjoining them submerged, and where this work is properly carried out it results in considerable benefit to the fishing of such pieces of water. The Loch Leven of the present day is too well known to need much description; suffice it to say, that owing to the praiseworthy efforts of those who have it in charge, a large and valuable stock of fish is maintained by means of artificial cultivation. The pike, which used to be plentiful in the loch, are also well attended to, no pains being spared to reduce the numbers of these voracious fish.

As far as regards our fresh waters, at least, there is no doubt that when properly cultivated they are valuable, and there is great room for making improvements in this direction. There are, indeed, so many instances in which the benefits of aquaculture are plainly seen, that the fact needs but little demonstration. Where a little has been already done much more can be done; the great desideratum is the knowledge as to how to do it, and this highly important part of the subject I shall endeavour to deal with pretty fully in the second part of this book.

There is another "Angler's Paradise" which is, I trust, on the verge of great improvement, and that is the English Lake District. Knowing it as I do, and having spent a goodly portion of my life there, I have perhaps a right to speak with some authority on a subject in which I have taken the deepest interest. The natural advantages and resources which this district possesses are difficult to estimate. The only question that can be asked is: how are they to be developed to the best advantage? This is the great question which has to be faced in attempting to deal with these resources. Where so many fisheries occur in such close juxtaposition with each other, as they often do in this district, it requires a clear course and a good deal of forethought, based upon a knowledge of locality, and above all, of practical fish culture of the most advanced type, to enable any individual, or committee of individuals, to deal with them in a satisfactory manner.

Take the case of Windermere for instance, and we have fronting us at once : the salmon fishery, the char fishery, the trout fishery, and the coarse fish, including eels. In these matters alone we find ample work for a local committee. That great improvements can be made there is no doubt, and I, for one, long to see the work well in hand, for a more hopeful case can hardly be. There is the raw material to work upon in almost unlimited quantity, and what a boon it would be, not only to the inhabitants of the district, but to the crowds of visitors who go there to spend their holidays, if the waters of this lovely country were really well stocked with fish.

Most of the rivers and becks of the district are very rapid streams, with essentially stony bottoms, and the water beautifully clear and transparent, whilst some are browned by the peat, amongst which they have their origin. Some, I regret to say, are poisoned by the foul matter sent down from mines or manufactories. The unpolluted streams are, in very many cases, exceedingly adaptable, and the development of such as these is a work of no mean kind. Many a beck, that now produces nothing but the finest water and a few small trout, may be made a driving power for scores of miniature lakes which would be controllable. This is exactly what some of the natural waters are not, although they too are capable of much improvement.

The char fishery is well worth developing, although the fish are not, perhaps, the most desirable from an angler's point of view. Their flavour, however, is so excellent, and they are such delicacies, that they might probably be made to pay for a part, at least, of the cost of developing the fisheries as a whole. It has been asserted that they will only live in deep lakes. True it is, that in such only they are usually found to occur, but in experimenting with these fish, I have reared them from the egg, up to a weight of three pounds, in a small concrete tank not exceeding three feet in depth.

Char are exceedingly sensible to changes of temperature, and probably, on this account, are found to inhabit deep water at certain times of the year. It is a well known fact, that one variety of char found in Windermere spawns in the river that feeds the lake. The fish migrate up this stream as the spawning time approaches. A short distance above the lake it forks, and the char invariably go up one fork only, and avoid the other. It is quite likely that temperature has something to do with this. Be it as it may, the char is a good fish, and its life history is well worth studying, with a view to its further development.

An idea seems to prevail amongst ichthyologists and others that these fish cannot live in the sea. I once kept one in sea water for twenty-one days, and it seemed all the better for it. The fish had got a little fungus *(Saprolegnia)* on the dorsal fin, and was put into a tank of sea water, to be taken out in a short time again. The man in charge of the work was suddenly called away, and the fish was overlooked for several hours, when I found it apparently quite happy in a corner of the tank. Seeing this, I allowed it to remain, but kept an eye on it, in case it became distressed; but nothing of the kind took place, and at the end of the twenty-first day it was taken out and returned to its pond, alive and well.

It is a question whether the introduction of a *Coregonus* (white fish) into some of the larger lakes might be advantageous. Where they are suited by their surroundings, some species are very prolific, and once introduced would soon multiply. These fish are known to occur in such numbers, in some lakes, as to form a very profitable source of income to those engaged in fishing for them. I am not yet prepared to say how far their introduction into such lakes as Windermere would be desirable; but the matter is well worth investigation. I have eaten them both in Europe and America, and have found them excellent. The cultivation of lake bottoms is a matter to which no attention has been paid, but it will be apparent to anyone that it must have a considerable bearing on the future welfare, or condition of our lake fisheries. There is a field for investigation here, in which there is room for a large number of workers—a piece of almost untrodden ground in fact.

In some of the waters of the district we find, on examination, a great variety, and a copious, supply, of the creatures on which trout and other fish feed, consisting largely of *Entomostraca*, whereas, in other places, there is a great scarcity of this minute yet exceedingly valuable fish food. The conditions that favour the growth of these creatures, and their presence or absence from any given piece of water, are matters of vital importance, and should be carefully studied. The insect life has also a very important bearing upon the future of the fisheries, and in some localities there is at times a great scarcity, although at other times there may be a great abundance of this food. We find there are two things which tend to stunt the growth of trout, and they are, uncertain supplies or scarcity of food, and low temperature.

The difficulty of low temperature can, however, be overcome, if on the other hand a good food supply is always available. I have grown fish to several pounds in weight in water of a low temperature by feeding them, and herein lies the main secret of success. Cold water and little food, which is the state of things existing in some mountain lakes or tarns, will dwarf the fish; but given a sufficiency of suitable food, the latter will improve. Where races have thus been permanently dwarfed for a long series of years, the introduction of new blood is undoubtedly advantageous, but the alteration of Nature's balance should be attended to first.

Apply the power, and so increase the food supply, the improvement of the fish will follow in the natural order of things. That many comparatively barren pieces of water might be so improved is certain, and this applies to any district where suitable trout water is to be found. Trout are increasing in numbers in Windermere, and in quality and size are excellent. This is what may be expected, for the pike have been successfully reduced in quantity, and a moderate amount of trout culture has been carried on. These two operations must tell their tale, other things of course being equal. There is scope in Windermere lake alone, and a natural food supply, that is capable of being made to produce very large results indeed. There are in Windermere, as in many of the other lakes, enormous quantities of eels, and these should be attended to as well as the pike, for they are most destructive to trout. A single eel, one pound in weight, will clean out hundreds of yearlings, or thousands of fry, in a comparatively short space of time, should it happen to succeed in gaining admission to a trout culturist's nursery ponds. The eel is one of our greatest enemies, but his depredations in natural waters are very much overlooked. When he gets on to a fish farm, however, he soon lessens the number of fish, if not destroyed.

A correspondent wrote to me this year, that he had lost a large percentage of his yearlings, owing to some eels getting into his ponds. I read in a book recently received from New Zealand, that "ninety rainbow trout were put into a large rearing box, with wire netting lids. These fish were growing splendidly, when some visitors to the ponds lifting one of the lids, and leaving it off, a large eel got into the box, and when discovered next morning its stomach was packed with nearly all this valuable fry, leaving only twelve alive." Such lessons as these should not be lost sight of.

During the last few years, the facility for visiting the magnificent scenery of the Lake District has been so much increased, that many parts of it are now very accessible to the tourist, and there is a great opening for the development of its waters, which did not before exist. No part of the world perhaps possesses so many charms for the contemplative mind. It would be difficult to find one which can provide so wide a field for the imaginations of the poet or for the legendary fancier, or such a charming variety of tint and landscape for the artist, as the lovely glens and varied hill-sides of this beautiful country. The lover of nature invariably finds much to delight him in this romantic region, and why, now that we have the power in our hands of dealing with the water, should it not be improved, so that it may be in the future more than ever it has been in the past, in the highest sense of the word—"An Angler's Paradise."

Part II.

HOW TO OBTAIN IT.

PART OF SOLWAY FISH FARM

HOW TO OBTAIN IT.

CHAPTER I.

FISH PONDS—CONSTRUCTION.

How not to make them—How to make them—Water supply—Sluices and over-flow—" Safety valve"—Leaf Screens—Ponds to be off the stream—Flood water kept out—Spawning beds—Barren water—Cultivation—Artificial Spates—Storage of water—Outlet screen—Effect of wind—Material for screens—Various kinds and importance of screens—Fontinalis rising to the fly—Bottom outlets—How to work them.

A N " Angler's Paradise " may now be placed within the reach of everyone who can handle a fishing rod. " How to obtain it " is the question before us. This I shall endeavour to answer, so far at least as to enable all who have the facilities, to commence operations forthwith. To enter into all details would be hardly within the bounds of possibility in so small a space. I must ask the reader, therefore, to pardon me if I appear not to refer with sufficient fulness to all matters connected with the subject. I will do my best to make matters plain and easy for all. Already many public waters exist where fishing may be had by anyone on fairly reasonable terms. There are other places also in course of formation, where the working-man can spend his Saturday afternoon or Bank holiday. It is now quite easy in many localities, by co-operation, for a working-man's club to provide excellent angling for the members at a very moderate cost.

In connection with the country house, mansion, or shooting lodge, a series of fish ponds will soon become in many cases a necessity. They need not in any way interfere with the five or ten mile walks up-stream or the delightful hours spent in the solitary glen or on the mountain loch. Rather will they become accessory or supplementary to such excursions, for they may be either in close proximity to the house and close together, or they may be scattered at intervals among the wildest of scenery; indeed they may be placed, so to speak, anyhow and anywhere, provided always, of course, that there is a suitable water supply.

To begin with, let us take an average country house. There is a small stream flowing somewhere not far off, and in it are to be found always a few small trout—a proof at once of its value, and the very best guarantee that can be as to its capabilities for supplying fish ponds. Don't on any account commence making ponds by damming up this stream at intervals, as has so often been done in the past. Such dams do not answer the purpose desired at all satisfactorily, and are often quite unworkable. Floods come down, and the result is that they get very much silted up, or filled with *débris* of some kind. The amount that is brought down by some of our mountain streams, when at times they become torrents, must be seen to be understood. Well do I remember remarking this to one of my early helpers as he was receiving instructions for making a dam, when with a knowing look he replied, "'Deed, sir; but there's a muckle heap o' 'deberis' comes doon here, when the water's oot!" This "heap o' deberis" (*débris*) must be reckoned upon, for sooner or later it is sure to come in most streams. In addition to the *débris*, which may consist of stones, sand, earth, or whatever the ground may be made of over which the stream flows, a good deal of other matter often comes down. I have seen ponds made by damming up streams that have given a great deal of trouble, simply owing to the quantity of floating matter that comes down during spates. Often the outlet has been made in the north-east corner of the pond and on the embankment. The natural consequence is that with every south-west gale, all the force of the water and floating material is thrown on to the outlet screen, whereas by a little forethought and arrangement this might often be easily avoided. The chances are that the screen is choked and the result is an overflow, by which probably some of the fish will be lost.

To make a fish pond, then, or a set of ponds, select any suitable place off the stream. It may be close alongside or it may be half-a-mile away. One of the best sets of working ponds I have ever made is about this distance from the point at which the water is drawn by means of a sluice from the main stream. They are on the side of a hill, and the fall from one to the other is thus excellent. The aqueduct passes through cuttings and along embankments, and where watercourses of a dangerous nature cross its track they are carried over it by means of wooden shoots. The sluice box in this particular instance is two feet by one foot, and under no circumstances can more water be drawn from the stream than can pass through this box. A much smaller one than this, however, or a pipe, may often be used with great success. I have seen a series of ponds working very well that were supplied through a three-inch pipe.

At the lower end of a long open aqueduct it is sometimes desirable to have another sluice and an overflow for waste water, as during heavy rains a good deal of surface water necessarily comes down. And here allow me to give a word of caution. Avoid lessening the regular supply from the stream at a time like this, or the water passing into the pond will contain too much of this surface water. Ponds supplied in such a manner, if properly made, are entirely under control and thoroughly workable. Too much water cannot get into them. On the contrary, during flood-time the quantity is found to lessen, as it is desirable to have a leaf screen in front of the sluice, and the flow of water is reduced by the mass of leaves or other floating matter brought down at such a time. I have found this arrangement work most satisfactorily—it is in fact the safety valve, and without such an arrangement in some form or other a pond or series of ponds is seldom safe.

There are several ways of constructing a leaf screen. The simplest of all for the present purpose is to drive suitable wooden stakes into the bottom of the aqueduct, where the soil will allow of it. I make them of larch, and cut them three-sided rather than square, and about one to one and a half inches in diameter. Drive them all carefully so that the apex of the triangle formed by the three sides of each stake is up-stream, and then place a rail across their upper ends behind them, and nail each one to it. Should any of the ends stand higher than the rail, owing to not having been all driven the same distance into the ground, take a saw and cut them off. The rail may be nailed to posts driven into the bank at either end, or otherwise made secure according to circumstances. This screen has a very important part to play, and if properly fixed, will do it well. Should the bottom be unsuitable for driving these stakes, they must be fastened to a frame, and while the water is temporarily stopped, this frame may be fixed in concrete, which holds it firmly and prevents the water undermining it, which it is very likely to do if not made secure.

Below this leaf screen is the sluice, which is a very simple contrivance, and below it again is a screen for preventing the escape of fish. This screen may be made of wood or iron. A wooden frame covered with perforated zinc answers very well and is easily renewed, but masonry and perforated iron plates are more lasting. Ordinary perforated zinc has to be renewed once a year. In some positions it is desirable to have what we call a " horizontal leaf screen." The term " horizontal " is perhaps not absolutely correct, as the screen is really placed at a slight angle, but is usually called a " horizontal screen " in contra-distinction from the " perpendicular screen " which I have already described.

A horizontal leaf screen consists of a box wihout a lid, sunk in the bed of a stream, and covered with perforated metal or a grating where the lid would otherwise lie. The box should be sunk so that its perforated top is even with the bottom of the stream, and has an easy gradient which will prevent its becoming easily choked. The water supply is taken from the box, underneath the perforated cover or grating, by means of a pipe. Instead of a box, the whole may be made of concrete if preferred. This screen will be more fully described later on, and if made use of here will prove to be an impassable barrier to the fish, and will do away with any necessity for a sluice or a second screen. Sometimes it is found to be easier, instead of making ponds off the stream, to divert the course of the stream itself, as for instance in a narrow valley. It amounts to pretty much the same in the end, being simply a case of keeping the main stream off the ponds, instead of the ponds off the stream. The object, of course, is to get rid of all flood water and·to have the supply for the ponds thoroughly under control. A reference to the accompanying diagram (Fig. 1) will, I think, make the arrangement sufficiently clear. Let AA be an embankment and KK a pond or dam above it. BB is the old bed of the stream, the course of which is now

diverted into the channel DD.   The water flows from the direction
shown by the arrow at F, and what is not required for the pond
passes off by the channel DD, which rejoins the stream below the
dam.   At the point *d* a horizontal leaf screen may be placed in
the stream, which passes over it, and the amount of water required
for the pond KK is drawn off from the box underneath this
screen.   Should a perpendicular leaf screen only be used, then let
it be placed across the stream at the point C.   A sluice may be
fixed just below it for greater safety, but my experience of the
working of these screens is that no sluice is as a rule required, as
the surplus amount of water gets away by the channel DD, and

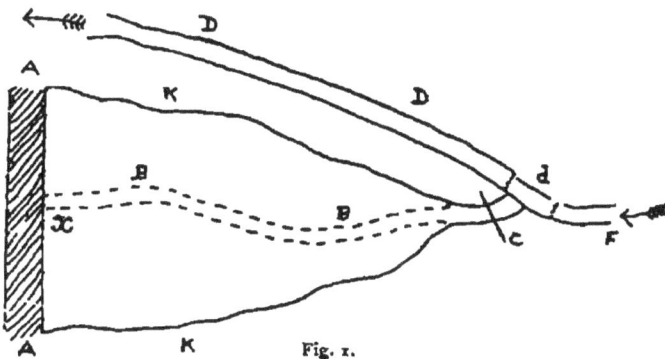

Fig. 1.

the screen at C becomes partially stopped by floating matter and
prevents any excess of water entering the pond.   One great
advantage of the horizontal screen is that no fish can escape from
the pond in that direction.   Where a perpendicular screen only
is made use of, it is needful to have another screen below it to
keep in the fish.

It is not at all necessary that the leaf screen be close to the
dam.   It may be some distance from it, so as to allow the fish
access to some spawning ground, and this spawning ground may
often be largely increased by making a serpentine spawning race.
This, in a great many places which I have examined, the ground
will admit of.

Care should be taken to note the highest flood level of the water, and to have the stream banked at the points F and *d* and along DD, so that no flood can possibly cause an overflow into the pond. This is a matter of the utmost importance, and one which I find is often overlooked, or for which sufficient provision is not made. The highest point reached by the greatest flood ever known should be ascertained, and the bank raised above that level. I have been surprised to find how careless or thoughtless many people are in guarding against floods, or in having ponds or artificial hatching beds in such a position that they are in great danger of being washed away. They may work well for one year, or two or three, when, owing to a combination of circumstances, an unusually heavy flood occurs and disaster is the result. I cannot remember a single instance that has come under my notice in which the danger might not easily have been avoided from the first. So there is no excuse for newly-constructed fish ponds being flooded.

In guarding the ponds against floods, it will be seen at a glance that another great point is gained, and that is, that the spawning ground is also protected. There is no longer any fear of the eggs being washed away, nor yet buried hopelessly, as they often are in our natural streams when the floods come down. By placing the leaf screen high enough up-stream, that is, some distance from the head of the pond, very good provision may be made for the fish depositing their ova in artificial spawning beds, where the eggs can be protected from at least some of their enemies. When the early fish have spawned, a grating or screen may be placed at the mouth of the stream to prevent any later fish coming into the beds, and rooting up the earlier and more valuable eggs which have already been deposited. This is a matter which has been grievously neglected in our streams and rivers, and which ought to have immediate attention bestowed upon it. I have seen whole beds of valuable ova turned up and the eggs devoured, and a lot left in their places which are not only of far less value, but are calculated, as time goes on, seriously to depreciate the stock in the stream. In the case of a salmon river it is a serious matter.

Where there is room at the head of a pond or a lake for the

construction of these protected spawning beds on a fairly extensive scale, they may be divided into sections, and the fish allowed access to the various sections as they spawn in succession. The first fish that come up, for instance, will push forward until further progress is barred by the leaf screen. Finding they cannot get any higher up-stream, and at the same time finding suitable spawning ground, they will deposit their ova, and return to the pond from whence they came.

Now, by placing a screen or grating below this spawning ground to prevent more fish coming up on to these nests, the later fish are obliged to deposit their eggs lower down, and when the beds are all made use of the fish should not be allowed to enter the stream at all. The latest fish will then spawn in the pond where the stream enters it. It may sometimes happen that a pair of the earliest fish, instead of pushing up-stream, will spawn in a lower reach of the raceway which is wanted for the later fish. Such an occurrence cannot always be avoided, but such a nest can be protected by wire netting from the depredations of the later fish. Or, if observed in time, the fish may be driven off, or netted out and carried higher up-stream. Any one who is determined to carry out the work thoroughly can easily do it; but in such a case it will require a little attention. It possesses one very satisfactory feature, however, and that is that it will amply repay the amount of care and attention bestowed upon it.

And now a few words as to the preparation of the spawning ground. By altering the natural course of the stream and making it wind back and forward, a considerable amount of ground may often be made available, and it is well worth while to take some pains with this part of the work. The great point is to make as much available spawning ground as possible. Being beyond all danger of flooding there is no fear of the work being washed up or half destroyed by a spate.

The stream which is to be made into a spawning race should have an easy gradient, so that the water just ripples along nicely over it, and the bottom should be of gravel. This gravel bed should be from six inches to a foot deep, and should consist of stones, about one third of which may be about the size of walnuts,

one third the size of hazel nuts, and one third the size of peas.
The two first may be mixed together and laid at the bottom, the
smaller stones forming an upper layer.   Over this may be
scattered, so as to form just a covering, a very thin layer of
material, which may be described as fine gravel or very coarse
sand.   Care must be taken, however, to keep out of the raceway
anything of the nature of fine sand.

In width this raceway should be about three feet, and should
the natural gradient be too steep, a fall, or perhaps more than
one, should be made, so as to bring the part available for
spawning to a proper level.   The best place for this fall is at the
leaf screen, if it can be so arranged, or at the point where the
water enters the pond.   By making it at either of these places the
spawning race is not so much interfered with, but, of course,
much must depend upon the nature of the ground.   This being
done the gravel should be put in.   Then a series of very low
weirs should be made.   They need not stand more than six
inches above the water level, but the depth of the water itself
in the reaches formed by these obstructions should run from six
inches to a foot.   Whilst the raceway is being prepared the water
should be shut off so that the work may be properly carried out.
The reaches of water may be ten feet or ten yards in length,
according to circumstances.   The divisions between each are
best made of stones or blocks of concrete.   I have seen very
effective ones made by simply placing logs across the stream, and
where plenty are at hand and stones are scarce they do very well.
Having formed these miniature dams and made all right, the next
thing is to turn on the water and wait results.   The effect of the
current will be to wash out holes below the weirs, and probably
to form a bank just beyond each of these holes.   One great
advantage of having the water under control is now apparent, and
that is that the bottom of the raceway can be raked over, and the
gravel spread to suit circumstances.   These banks may be raked
down, and the gravelly bottom otherwise improved, if found
desirable, by hollowing it out a little in the middle.   When the
fish come up they will soon make short work of it, however,
rooting it up everywhere in making their nests.

At the spawning time especially, as well as at other times, it is desirable to run a good stream through a pond. I have seen many worked most successfully where only a very small water supply was available. These are cases of necessity, however, and although in many of them good results may be obtained, yet it is only by judicious working that they are brought about. By limiting the quantity of fish, paying special attention to the existing conditions, and regulating the work, much may often be successfully done.

There are cases in which fish ponds have been made on cold barren streams, that have received their filling from torrents of snow water or cold winter rains, and even before the workmen have been off the scene a quantity of yearlings or two-year-olds have been hurriedly introduced to "save a season." Sometimes these fish, if not too numerous, have done very well for a while, but it has been more by accident than otherwise. I have known cases in which they have done badly, and it is not at all surprising that it should be so. It matters not where the fish come from or how good they may be, the result will be the same. Anyone who understands the keeping of an aquarium will readily see wherein the mistake lies. Before turning fish into a pond we should know that there is food upon which they can not only live, but thrive ; but in many cases which have come under my notice little attention has been paid to this important matter. The water has been stocked regardless of surrounding circumstances, and consequently expectations have not always been realised. Although grave errors have been committed in the past, I would not for a moment blame anyone. Even fish culturists have been years finding out all these things, and there is yet very much to learn. But we are now fairly on the right path, and mysteries are being rapidly unravelled

We are now possessed of sufficient information to enable us to handle the water in a practical manner, as the agriculturist does his land. As there are many varieties of soils, and as these require different courses of treatment, so there are different classes of water, each requiring its own peculiar management to make it capable of producing a crop. As a rule, however, the cultivation of ponds and lakes is a much easier matter than the cultivation of a tract of land. Where the water supply is small it is very

desirable, after filling the pond, to cultivate it properly, and to
bring it into a suitable condition for receiving the stock of fish
that is to inhabit it. As a farmer must provide a supply of food
for his sheep before turning them into a field, so we must provide
for our fish.

Trout are carnivorous, and if we examine the contents of
their stomachs we shall find that the creatures on which they have
been largely feeding are mainly dependent upon a crop of
vegetable food. If they are to exist in sufficient numbers to form
an adequate supply of food for the fish that prey upon them this
must be provided. Therefore, it follows that by judiciously
cultivating suitable plants a sufficient food supply can be maintained, other things of course being equal. Should this food
supply not be in existence naturally in the water it must be
introduced, and regarding its introduction, and the introduction
of aquatic vegetation, I shall have more to say later.

Where the water supply is of a limited nature, it will do
better work and go further in a deep pond than in a shallow one.
It is obvious that in a pond of considerable area, but only running
three feet or so in depth, the water will be more affected by the
rays of the sun, and in summer will rise to a higher temperature
than in a pond of smaller area but which is eight or ten feet deep,
or perhaps more. The pond of smaller area but greater depth,
will support a larger quantity of trout during such a period than
the shallow one.

It also often happens that in addition to the visible water
supply running into a pond there are springs in the pond itself, or
there is a considerable quantity of water percolating through the
soil, and this alone is a very valuable help to its trout producing
capacities. In the deeper pond there is a greater probability of
such an additional supply. I have known ponds which have thus
been fed by copious springs, but which have possessed no visible
water supply, that is to say, there was no stream running into
them, and yet they have proved capable of maintaining a nice
head of trout. Sometimes we find a pond from which there is a
considerable overflow, but no stream apparently feeding it. In
such a case there must be springs, and such a pond may do for
trout. Occasionally in ponds of this nature, however, the water

is too cold or barren, and will not grow trout at all, whereas, the same water supplying a second pond will do very well for the purpose. The fact is, that it simply requires a little cultivation to render it fit for use. After being exposed to the air and the rays of the sun in the spring pond, which contains also some vegetable life, it becomes changed in its nature, and in passing out of this pond and along a raceway, even though the latter be only a few yards in length, it becomes still further changed. This applies more or less to a great many springs. The water often becomes so rapidly changed that trout will thrive even in a spring pond.

I have had cases in which they have done well in ponds that had neither visible inlet nor overflow. Being fed by springs the water would rise to a certain level, when, owing to the nature of the soil, the absorption was so great that it could rise no higher. There was, in fact, a stream flowing through the pond, and a pond of this nature may be an excellent place for trout.

Water readily takes up oxygen, and the more rapidly this combination is brought about the sooner does it become suitable for maintaining them. Whilst receiving a supply of oxygen it also precipitates substances held in solution, and thus it will be seen that the water flowing out of a pond may essentially differ from the water flowing into it, if it be from springs. Plants give off oxygen, and, as a rule, in other ways tend to render it more suitable for fish. It is a well-known fact that even impure water as in the case of a stream, becomes purified as it pursues its course. Exposure to the air and the absorbing power of the earth and other causes are at work to bring about this change. Where the water supply is obtained from springs, therefore, it is desirable to pass it through a preparatory pond if practicable, and it is important that this pond should contain vegetation of a proper kind, as the plants not only give off oxygen but produce favourable conditions for a crop of natural fish food. So much, then, for ponds with a limited water supply.

Where a good supply is at hand I would say secure it by all means. A stream that will fill a twelve-inch pipe will keep a good series of trout ponds going. If double the supply can be obtained at times so much the better. Although it is not only desirable, but necessary, in the case of a successful trout pond, to regulate the water, yet the supply should not always be just the same.   It is not so in nature.   There are freshests or spates which are often very beneficial to the fish.   Now, we have the advantage in a well ordered fish pond of producing these artificially, whenever desirable, and what is more, we can have the spate without a flood and without fear of an overflow.   Anyone who has seen an artificial spate sent through a pond during dry hot weather in July, would never doubt again the beneficial influence it had on the trout.   We know that in streams they often suffer very much during droughts, and occasionally even die in numbers.   What a great advantage it would be in such a case to send down an artificial spate occasionally, giving the fish not only additional water, but taking care that that water was charged with trout food. It could easily be done and would be an invaluable help, and I venture to say it will be done before long.

I shall never forget a case in which I was once called suddenly in the month of July, to look at some trout in a pond that had been almost deprived of its water supply during a long drought.   When I arrived I found but a mere trickle going into the pond, where the sun's rays had heated it to a dangerous degree.   Not a drop was running out, and the water had fallen considerably below the overflow level, owing to the loss by evaporation and soakage.   The trout were gasping on the surface, and a crowd of them also in the same condition were about the inlet.   The case looked hopeless, but *nil desperandum.*   The pond was supplied from a mountain stream, and this stream, like hundreds of its class, was almost dry.   But it consisted, as most mountain streams do, of pool after pool, and some of them were of fair size.   Obtaining a couple of men armed with crowbars, spades, and pickaxes, I had them at work in a few minutes letting off these natural water supplies.   There was no difficulty about it, as the stream bed consisted mainly of boulders and gravel, and by shovelling away a little of the latter or lifting a boulder or two a considerable amount of water was liberated, and in a few minutes a small artificial spate was going merrily through the pond.   The fish revived at once, and thus a valuable stock was saved from certain destruction.   Had this not been done, at least three-fourths of them, if not all, would have been dead in a few hours. As soon as the water in the pond had become thoroughly changed I allowed the stream to slacken considerably, and during the night it only ran as I had at first found it. The fish had recovered, and water was valuable, as we did not know for how many days the drought might continue. Next day, however, the operation of letting off more pools was continued, although with great care, no more water being sent down than was absolutely needful to keep the fish in a fairly good condition. This operation was continued for five days more, when a friendly thunderstorm drenched the hills and sent down a copious supply, which did not fail again that summer. Before the next a supply had been obtained from another and larger stream.

In the instance I have just narrated the fish were not only benefited, but their lives were actually saved, by an artificial spate, and the great benefit arising therefrom will be plainly perceptible. So at any time during dry weather, although there may be no danger whatever existing, an artificial spate sent through a pond is an exceedingly good thing for the fish. It may be produced by contrivances of the roughest and simplest kind. A few small dams should be made, not across the stream itself, but in any suitable corners close to it, where a trickle of water is obtainable to fill them. This means that there will be practically hardly any stream going through them, and thus they become excellent places for growing a crop of trout food, which, of course, goes down with the water into the fish ponds when the dams are let off. And here it should be observed that these dams should never be run dry. By only running half the water off, or at most three-fourths, a sufficient stock of living food is retained in them to keep up the required supply.

Having made secure the point at which the water enters the pond, the next thing is to provide for its escape when the pond is full. There are many methods by which this may be effected. One that I have found to be in very general use, is to allow it to take the shortest possible cut over the embankment, in a channel the bed and sides of which are more or less paved. They may, of course, be built of solid masonry. So far good; but there are many things to consider, which I find in most cases have not been thought of at all. Should the pond be in an exposed situation, it is often desirable to make the outlet as near the south-west corner as practicable. Our prevailing wind being south-west, and most of our heavy gales coming from that quarter, it will be apparent that the effects of the wind will be to keep the south-west corner clear of floating matter. When the outlet is in the opposite corner, all the floating matter is driven right on to the screen, and causes much extra trouble. In some cases the situation of the outlet in a rather bad position as regards the wind is almost unavoidable, but in a great many instances it is quite a simple matter to arrange it on either the southern or western side of a pond. Of course some pools are sheltered on the south-west, and exposed to east winds, and work best with the outlet on the east side. It is therefore necessary to take the bearings and study the peculiar circumstances surrounding each case, and then decide on the best place in which to construct an outlet. The result may perhaps be that the best position for it, as regards wind, may prove to be on the same side or very near the place at which the water enters. This, again, is not good, and it may be desirable to convey it, by means of a raceway, round to the opposite side, and turn it in there, so as to make sure that it runs completely through the pond. This may often be easily done, and is an advantage, inasmuch as it provides an extra length of spawning ground for the fish. Or it may be done by placing the leaf screen half-way along the raceway DD (see Fig. 1) instead of at the point C. or D. So much, then, for the position of the outlet.

Here let me say that the east wind is sometimes as bad for fish as it is for man and beast. It has been found that under some circumstances fish are adversely affected by it. This is one of the many points that require investigating, and every scrap of information that can be gathered together will tend to teach us something more about that of which we know comparatively little at present. We do know that in laying out fish ponds for profit, which are necessarily small and heavily stocked with fish, the east wind is a factor which has not to be overlooked.

Having considered the position of the outlet, we have now to decide which is the best plan to make one—one that can be relied upon to work in all weathers, with the least possible attention or care on the part of the attendant. That care must be given to an outlet screen goes without saying. It should be the duty of some one to attend to it as often as may be required. The amount of attention needful varies very much indeed in different cases. I have screens working which are not touched oftener than once in three or four weeks, and I have others that require attention every day and sometimes twice daily. There are exceptional times, that come perhaps once a year, when a screen requires special attention for a few days, as, for instance, during a slight rise in the water in autumn when the leaves are falling. As I have already pointed out, a good deal depends upon the position of a screen, and much also upon the way in which it is guarded. I have seen one that was almost hopelessly unworkable work quite easily after steps had been taken for preventing the floating matter from coming on to it, and this can often be done in a very simple manner. A few bundles of sticks or of thorns placed in the water so as to form a semicircle just above the screen have done excellent service many a time. A wooden hack placed two or three feet in front of it will answer the same purpose. I have seen wire netting used, but it has the objection of being bad to clean. From a wooden hack, or an iron one either, rubbish can easily be raked, but not so easily from wire netting. Should it be the most convenient device at hand, it should be stretched upon movable frames that can be lifted bodily and well shaken, or the netting beaten with sticks or switches to clear it of the mass of material that is at times sure to collect upon it. Bundles of thorns, hedge clippings in fact, do very well, and if plenty of them are used they are most efficacious.

The working of a screen may also be materially helped by planting a bed of reeds (*Arundo phragmitis*) or of bullrushes (*Typha latifolia*) in front of it. These will do excellent service in keeping it clear, by preventing the bulk of the floating matter from reaching it at all, and by working a few bundles of sticks at the back of a reed bed a screen may be made to give very little trouble indeed. "Where there is a will there is a way," but the disgraceful manner in which many of the outlet screens I have seen have been worked would lead one to suppose that it was a very difficult way. On newly-made fish ponds, however, it should not be so.

F

The next question is—What is the best kind of screen? There are several important points to notice.

(1) A screen should fit well into its place.

(2) It should be strong.

(3) It should be well let into the bank at each side, and especially underneath, so that the water cannot by any means flow round it or underneath it.

(4) It should be large enough.

(5) It should be so constructed that it will not allow any fish to escape.

Only on very rare occasions have I come across screens that possessed all these qualifications. In travelling about the country inspecting ponds and lakes, I have not found one piece of water in fifty to be possessed of such a screen. Many have been introduced at my suggestion, and have worked most successfully. The proper working of a screen depends largely upon the man in charge of it. Occasionally, one meets with an individual whose mind is made up before he sees it or begins to attend to it, and in such hands it has usually a poor chance. I have seen one that would not work at all under the care of one man do its work admirably when in charge of his successor.

There are several different forms of screen which we will consider by and by. Before doing this, however, let us turn our thoughts to the material of which they are to be made. I have at present close upon a hundred screens working about my ponds, and I have always used perforated zinc, and have found it to work well. It is fitted on to wooden frames, and these being all of the same size, one can be withdrawn, and another of coarser or finer perforation be readily put in its place. This on a fish farm is very necessary, as large fish may be taken from a pond and replaced by small ones needing a finer screen. Perforated zinc works well, is not expensive, and is easily manipulated by any one who can use a hammer and nails. Should a hole get punched in it by accident, it is easily repaired by lacing another piece over the hole with the help of a little brass or copper wire. Ordinary zinc requires renewing about once a year, and this is its only drawback, but to set against it is the advantage it possesses in being so easily put on. It can be obtained of considerable thickness, however, and then lasts much longer. Perforated iron plates do very well, and are of course more permanent. They require no frames, and are quite easily fixed. An ordinary iron grating is also excellent, but it must be borne in mind that a fish can readily pass between two bars of metal which could not possibly pass through a round hole of the same diameter. The bars may be a quarter of an inch or they may be half an inch apart, but the greater the width between them the more readily will fish pass through and escape. It may be argued that fish do not go down stream out of ponds. I know many ponds where there is no outlet screen, and where it has been supposed there is no need for one. This is a mistake, however, and I have met with instances in which fish have been lost in considerable numbers owing to it. At times, and under certain circumstances, they will and do go down stream if not prevented. There are some instances, however, in which a large number of fish may run up stream into a pond, and of course all things have to be taken into account, and each particular case treated according to circumstances. One thing I am quite sure of, and that is, that every fish pond that has been specially made and stocked with fish, and that has a regulated supply of water, ought to have an outlet screen.

Having decided upon the material that is to be used, the next thing is to consider the more important points that are to be observed in the fixing of a screen. We must first be sure that it is capable of passing all the water during rainy weather. This is in itself a very simple consideration, and in the case of a well made pond that is not subject to flooding, is easily calculated. By having a screen that will pass double the quantity of water that can by any possibility flow from the pond, it should be fairly safe. Of course, the amount of floating matter that is liable to be driven on to it must be well considered and allowed for, and it is better to have the screen twice as large as is necessary than to have it just a little too small. One double the capacity required is usually large enough for all emergencies.

There are so many old ponds in use that are liable to have floods tearing through them, that I will just give a hint or two concerning them. The chief point in their case is to have the screen large enough. Make it as big as you like, but do not on any account have it too small. It will do no harm by being made too big, and will give less trouble. Where the size can be given in width and depth the matter is simple, but sometimes this is not convenient. There may be an opening through which the water passes that cannot be easily altered, and which at the same time is too narrow for a screen to work. In such a case the screen must project into the pond, and may be on the rectangular system—that is, consisting of two sides and a front—or it may be made semi-circular, and in this case may be much wider than the outlet. In cases where there is much floating matter and heavy flooding, it may be commenced several yards from the outlet, beginning on the pond bank, and running out in semi-circular form until it reaches the bank again, a few yards on the other side of the outlet. It will be apparent that the greater the dimensions of the screen the less will be the liability to become choked, and, consequently, the more easily will it be kept in working order.

It is very desirable that a screen should be well fixed, that is, that it should be so let into the banks of the pond that there is no chance of the water flowing round it, or under it, instead of through it. This end can be attained by means of masonry or concrete, in which the screen may work, and which must be well puddled behind. A very simple and effective way of fixing a screen, when made of wood and perforated or woven metal, and one which I have frequently followed, is explained by the annexed diagram :—

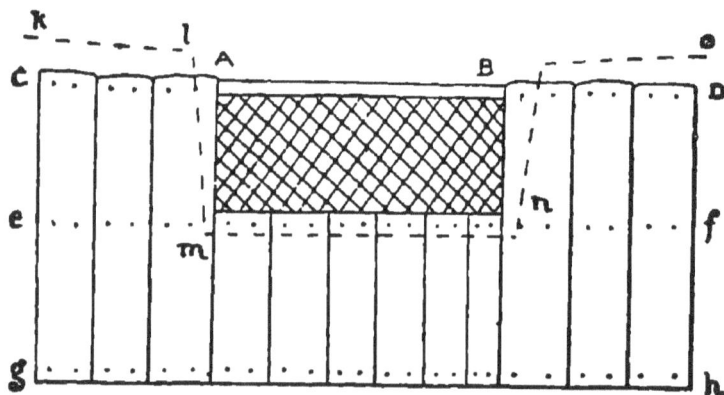

Fig. 2.

A B is a bar of wood, underneath which is the wire or perforated screen, as shewn. On the outer ends of this bar, where it extends to C and D, are nailed several boards. Another cross bar lies at the back of the boards just below the screen, *e* to *f,* and another at the bottom of all, *g* to *h.* The whole of the wood-work below the dotted line *k, l, m, n, o* is to be buried in the bottom and banks of the raceway, and, in constructing it, the screen should be made of a suitable size to fit the raceway, unless it be intended to use a projecting one. In such a case no screen is fitted into the opening below A B, but the projecting one may be fixed to the woodwork, and instead of working in the raceway it works in the pond. There is a great advantage in using wood, inasmuch as joints, alterations, and repairs are easily made. It is true that it is more liable to decay than stone, but it lasts a long time, and is not so often damaged by frost, and it has the advantage of being less expensive and more easily adapted than masonry. It should always be charred, both for the sake of preservation, and to prevent the growth of fungus *(Saprolegnia).* A screen constructed in this way will be found a very simple affair, and the filling of the trench which has been dug to receive it, if properly done, will render it perfectly tight and secure. If the work be done in concrete or masonry the principle is still the same, and will ensure safety unless it has been very clumsily managed indeed. The screen can be made to slide in a groove or be a fixture, as may be most desired.

There is not much fear of the water getting round or under-neath such a screen, and it will be found to be worth the labour expended upon it. Fully three-fourths of those I have come across which have been made in some other way have been found to be useless. Once I was called to inspect a small lake that had been stocked by an enterprising hotel proprietor. The fishing had not improved as it should have done. We made a careful survey of the water and its surroundings, and on approaching the outlet one of the causes of failure was very apparent. The outlet screen consisted of a big wooden frame with perpendicular iron bars, but it was choked up so that no water could pass through it, and instead of doing so it was escaping through a hole which it had washed out round the end of the frame, forming an aperture through which I could almost have crept. When I pointed this out to Mr. Blank and told him that most of his fish had bolted, he threw up his hands in astonishment that such a thing could be. What else could have been expected from a neglected screen, a rat hole and the action of the water in flowing through it, but the result I have described?

Similar occurrences I have found to be very common—water escaping from a pond otherwise than through the outlet screens. Sometimes in a paved raceway it takes the form of bubbling up here and there among the stones, or of coming out at a hole in the side of the raceway some distance from the pond. These things ought not to be, and wherever they do exist they should be remedied in the near future. What would be thought of a farmer who put his sheep into a field, but did not shut the gate? So long as it served the purpose of the sheep to stay in the field they would do so, but no longer. It is the same with fish. There is a widespread belief that when they are put into a pond they will stay there, but my experience is that in a great many cases they do not do so. A considerable number of them may remain, perhaps enough to keep the pond always stocked, but if the best fish be disappearing every time there is an inducement to them to go, how much better it would be to prevent all possibility of such an occurrence by having a screen at the outlet. With such a provision and a good spawning ground, a well-stocked pond or lake ought to give very good results.

Many ponds have been made during the past few years, more or less in accordance with the instructions which I have given, and where the work has been properly carried out the result has been most gratifying. I will refer only to three of these ponds which were made by R. A. Yerburgh, Esq., M.P., at Barwhillanty, in the County of Kirkcudbright. The first pond was finished in the year 1888, and was stocked with trout fry on May 14th of that year, according to my instructions. The fish grew splendidly, and two months later were seen rising to the fly in all directions. So decided was the success of the undertaking that a second pond was at once commenced, and a good staff of workmen being employed it was speedily completed, and was stocked the following year, January, 1889, with large fish. A sufficient supply of natural food had been carefully prepared, and the water was in thorough condition when the fish were turned in, the result being some excellent fishing during the following summer. A third pond was then constructed and was stocked with yearlings, which also did well. The fish introduced were yellow trout, Loch Leven trout, and American trout, and all gave excellent results.

The vexed question as to *S. fontinalis* rising to the fly was settled here beyond a doubt in July, 1890, by G. Ward, Esq., of Lake Vyrnwy. Writing to the *Fishing Gazette* of July 26th, 1890, Mr. Ward says :—

"Dear Sir,— I cannot allow a most interesting visit I have just paid to Mr. Armistead's fish hatchery and breeding waters to pass over without giving you some particulars of same. A rod was given me, as a special favour to cure me of my doubts as to *fontinalis* rising to the fly, and at the first cast that question was decided. I caught six in a few moments, and the way these trout rose was certainly 'a caution,' as in a special pond, kept for the large fish, two rose at one cast and took both my flies, I having a stiff rod that was too 'hard' on the fish. Mr. A. now suggested that I should go to a water he had stocked three years ago, which was a few miles from his place, and, nothing loth, we started to see fresh wonders. On arriving we made requisite arrangements . . . . and with very coarse and clumsy tackle I commenced to cast ; but it made no difference to these fish, especially the *fontinalis*, who rose to me in a most dashing manner, although the water was not rippled. I had more than a dozen landed in about half-an-hour, all of which required playing, twice having two fish on at once, and I may say I never saw anything like it in my life. . . . . Mr. Armistead could have taken me to a number of other places where similar results have been obtained, but I had seen enough. I must conclude by thanking you very much for having, by your kind introduction, given me the opportunity of seeing so much to interest the sportsman and fish culturist.

Yours very truly,

"Lake Vyrnwy."   G. WARD."

The last interesting experiment was allowed by the kind permission of the factor, and I need hardly say that the fish were carefully returned to the water.

Every fish pond should have a bottom outlet, by means of which it can be run dry at any time that may be desired. It should always be borne in mind, however, that when the water is let out of a pond the natural food contained in the water passes off with it, and that, on refilling, some time must elapse before this stock of food can be reproduced. Especially is this the case when the pond is supplied by means of springs, or the comparatively barren water of mountain or moorland streams. For this reason, a pond should never be let off unless for some very special purpose, as, for instance, when it may become needful to clean it out, or to change the stock of fish, or possibly to destroy pike or some other predacious fishes, which may have been introduced accidentally or otherwise.

There are many ways of constructing a bottom outlet, the great need being to have one that will allow the water to escape without at the same time letting go the fish. This outlet should be made at, or rather below, the level of the deepest part of the pond. In the case of a large reservoir, where there is much pressure, the pipes by which the water escapes should be of iron, and should be very strong; but for a small pond, or one where the water is not very deep, good sound earthenware pipes will often answer the purpose. I would here point out the necessity that exists, in the construction of embankments and bottom outlets connected with ponds of any size, for the employment of a thoroughly competent engineer. So much often depends upon the strength of a bank, that too much caution cannot be observed in attempting to impound any considerable amount of water.

I was once sent for to inspect a pond of some fifteen acres after it had been completed, with a view to advising as to the best course to pursue for getting up a good head of fish. I had never seen it before, and when I arrived I found it nearly full, and during my examination naturally asked a few questions as to the embankment, puddle trench, etc. This not being really part of my work, I did not say too much about it, but strongly advised caution, and the water was not allowed to rise to its highest level for some time, and then only very slowly. The bank was closely watched, and, as the filling went on, the water was discovered percolating through, and the owner at once ordered the bottom outlet to be opened. The outlet pipes were of earthenware and of considerable size, and as the foaming mass of water came pouring through them, it was observed that the pipes were breaking up and going away down stream. Fortunately the embankment stood, but it was a narrow escape, and I have thought it well to mention it here as a preventive of similar occurrences in the future.

The question as to which is the best kind of bottom outlet for a large reservoir, will soon be settled by the engineer, who should consult a fish culturist before deciding on so important a matter. I will simply describe one which I designed some thirty or more years ago, and which has in all cases worked perfectly, and is now in use in most of my own ponds. Where they are small it should be made of wood well charred. It is a very simple contrivance both in construction and manipulation, and consists of a wooden box about two or two and a half feet square, and half a foot or so deeper than the deepest part of the pond. Suppose this depth to be seven feet to the bottom of the outlet pipe, then the box should be made, say, thirty inches by thirty inches by seven and a half feet. It should be let into the embankment if practicable, but if, owing to the formation of the bank, it be a few feet out into the pond it does not really matter, except that it will require stays to keep it firm, and a gangway plank to connect it with the shore.

Now for the box itself. It will be seen at once that it is not the shape of an ordinary box, but may be described as long and narrow. There is also a little peculiarity about its make. Let us imagine it for a moment standing on one end, which we will call the bottom, and which should be very securely fixed on. It stands seven and a half feet high, and has four sides, one of which we will call the "back." This is to be placed nearest to the bank, and a round hole should be cut in this "back" as near to the bottom as practicable. This round hole should be just large enough to allow the end of the outlet pipe to enter and project a little way into the interior of the box. So much for the "back." The side opposite to it we will call the "front," and the others simply the two "sides." There is nothing special to observe about the two "sides" except that, like the "back," they are to be firmly put together. A good square corner piece or batten in each of the four corners is a great help to the whole construction, and, in any case, one must be fixed in each of the corners towards the front, to which we will now turn our attention. It faces the pond, and as yet is quite open, except a couple of wooden straps which have been fixed across it in order to keep the whole affair together and prevent collapse of the two "sides." The peculiarity about this "front" is, that the boards are not to be nailed on, but are to be quite loose. They are kept in position by each of their ends sliding into a groove which is formed to receive them, and being slipped into this groove one after another till all are in, it will be apparent that they close up the front side. A peg at each end of the uppermost board to keep it in its place, and a movable lid being made to cover up the top of the box, it is finished. The front boards will float at first when released, but a piece of lead nailed on each will settle the difficulty.

Before the box is thus closed up, however, a plug should be fitted into the outlet pipe, and let this plug have fixed into it an iron ring or eye, which must be placed so as to stand out horizontally, and be ready at any time to receive the end of a lever, by means of which the plug is drawn. This lever may consist of a piece of larch or ash, and is in reality a boat hook. It must have a piece of wood nailed on to its side at a suitable distance from the bottom end, to act as a fulcrum when being used. It should be fitted and tested before the water is let on, so that there can be no doubt about its action when required to draw the plug in an emergency. The whole contrivance is so extremely simple and answers all purposes so well, that I have never desired anything better. The box can be built of brick, stone, or concrete, and, working on the same principle, can be used for a pond of any size. In the case of a large reservoir the use of massive stonework is desirable, and iron or other plates in "front." Instead of working with a lever, the outlet pipe may be carried by means of a bend through the bottom of the box instead of the "back," and the ring in the plug will then be perpendicular, enabling the latter to be drawn by means of a boat hook.

Now, let us suppose the plug to be in, the pond full of water and fish, and let it be found needful to empty it. We go to the outlet armed with a lever, a long-handled garden rake, and a frame about three and a half feet square, on which is loosely stretched a piece of strong fine-meshed netting. Take off the lid of the outlet box, draw the pegs that keep the "front" boards secure, and lift out the two top boards. Having done this put the lever or boat hook down to the bottom of the box, and feel for the ring, which is quite easily found. Then insert the hook, and, taking care that the fulcrum rests against the back of the box, pull the lever landwards. The plug is instantly drawn and may be lifted out. There is now a rush of water over the top of the "front" boards, or as we call them on a fish farm "water boards." These boards do very well tongued and grooved, and I find six inches a very convenient width. By lifting two of them, therefore, the water is allowed to lower a foot, and the framed net is hung on in front of the opening, to prevent the escape of any fish. The pressure of the water will keep it in its place. When the water has lowered nearly a foot lift two more "water-boards" and lower the net, and repeat this until the pond is nearly empty, when the fish may be netted out and dealt with as may be desired. The garden rake will be found an excellent implement for lifting the lower water boards.

CHAPTER II.

Plants—Balance of life—Flora and fauna—Old ponds require cleaning—Pond life—Its bearing on fish life—Cultivation—Conditions of soil—Planting—New ponds—Virgin waters—Whitley reservoir—Importance of mollusca and crustaceans—Aquatic plants—Dalbeattie reservoir—Loch Fern—Plants to avoid—Weeding—Anacharis—Marginal plants.

M ANY people seem to think that all that fish require is plenty of water, but they must have a good supply of suitable food, or they will starve or become stunted in growth, like any other members of the animal kingdom. Food, be it remembered, is a necessity, and without a reasonable quantity of it, it is impossible to maintain a good head of trout. Fortunately we have the means within our reach of producing it in large quantities. The creatures on which trout chiefly feed are, directly or indirectly, dependent for their existence on vegetation, therefore it becomes absolutely necessary to have a proper selection of water plants present. There is undoubtedly a great future for some at least of our fisheries, but that future need not be looked for so long as such matters of importance are ignored. A farmer might just as well expect his land to provide handsomely without being cultivated or fertilized. Some natural waters on being stocked for the first time give results beyond expectation for awhile, but then follows a great falling off in their yield—like the virgin soil of the Western prairies, these waters soon become used up, so to speak.

It has been observed that where artificial ponds have been made by embanking, and the sod has not been removed, or only to a limited extent, that many of the grasses and other plants have not been immediately killed, but have survived for a period, and even for awhile become more luxurious in growth. While this has been taking place on the one hand decay has also been setting it on the other, and this decay has produced favourable conditions for vegetable growths peculiar to such situations. Trout fry turned into such ponds have developed into good fish, sometimes in an incredibly short space of time. The cause of this rapid development is to be found in the abundance of food produced by the surrounding conditions, but the submerged plants have at length decayed and disappeared, and with them, to a large extent, the parasitic forms for which they have provided a suitable *nidus.* As a result the trout have ceased to grow—nay, have even fallen off in condition; but aquatic vegetation having been introduced, in a few years the quality of the fish has improved. This needs little explanation, for the cause is obvious. Therefore, when ponds are constructed, a sufficient quantity of suitable vegetable life should be introduced. Some plants are peculiarly favourable for the development of fish food, some are positively hurtful, some are highly poisonous, and some actually devour fish. It will be apparent, therefore, that some knowledge of the subject is necessary before going to work, or the desired end may not be brought about.

A great deal more depends upon keeping up the exact balance of suitable vegetable and animal life than the uninitiated imagine; and I have seen considerable sums of money sunk to little purpose, in some cases thousands of pounds sterling, simply because the matter has not been understood, and consequently no attention has been paid to the most necessary details. It has, indeed, never occurred to many minds, that the utmost importance attaches to the presence or absence of various forms of vegetable and animal life in the water.

Before any serious expenditure is incurred on any lake or pond, the water and its inhabitants should be accurately studied. The results of dredging and the use of the tow net in various waters, goes to show that in some there is an abundance of trout food, whereas in others there is a great scarcity. Some contain many injurious pests to fish, whilst others are free from them. The study of *flora* and *fauna* is of the most vital importance; and the presence or absence of many microscopic as well as more conspicuous animals, too apt to be overlooked, will often render a piece of water fruitful or barren, as the case may be.　The more intimately we become acquainted with these important matters the more likely are we to manage them properly.

Many old ponds are so completely overgrown as to be choked with vegetation, and the only course that can be taken with most of these is to let off the water, clean them out thoroughly, allow them to remain dry for a period, and then deal with them as I shall direct presently.　The advantage of having a properly constructed bottom outlet, as described in my last chapter, will now be apparent, and at a time like this it will be found very useful.　At the same time it must be remembered that when the water is run off a large quantity of valuable fish food goes with it.　This stock of food has to be replaced before the water can again support the usual number of trout which the pond has been accustomed to hold.　When it is one of the lower ones of a series this does not so much matter, as water can be let down from the pond above, which is charged with animal life, and from this source a stock is soon produced.

In connection with the aqua-culture of the period, perhaps there is no subject of greater importance than the study of the various forms of life that are to be found in greater or lesser quantities in our waters.　To work out the life histories of the creatures, and ascertain the peculiar conditions under which each one exists or ceases to exist, and also those conditions which are most favourable for reproduction, in the largest quantities, of each species, is a work of no ordinary kind.　Passing over the more prominent and visible creatures, we come to a wonderful series of microscopic beings, which in some localities are found in enormous numbers, whilst in others they either do not occur at all, or if they do, it is simply a case of bare existence.　I have examined some waters in which the so-called lower forms of life, both animal and vegetable, are conspicuous by their absence, and the fish in these waters are very naturally stunted in growth, and are never known to attain any size, save where an individual develops cannibal propensities, and soon outgrows his fellows.　It will be seen, therefore, that the conditions under which fishes are wont to thrive are those to be first ascertained before stocking is resorted to.　It ought to be needless to point out that it is not enough merely to make a pond, and put fish therein and expect them to grow. It would be just as reasonable to fence in a piece of moor, turn down a lot of lambs, and without any further trouble, expect to receive in return a good crop of mutton. Practically, however, this is the way in which fish culture has often been attempted, and the result has been taken as a fair average of its value. Properly cultivated water will yield excellent results, but unless the right conditions exist, it is as impracticable to obtain them as is the case with badly cultivated land.

Take the case already referred to of a pond that has been neglected for many years, and that has in consequence become filled with weeds and mud. The first thing to be done is to clean it out, and having done this, to set about preparing the bottom for future use. The treatment necessarily varies according to circumstances. It should remain empty for at least a year, and during that time care should be taken that all noxious weeds are destroyed. The bottom should be cultivated. As soon as the soil of which it is composed is dry enough, plough it by all means, if it be at all practicable to do so. In a great many cases this can be done to a considerable extent. There may be places which remain soft and watery, and others where the bottom is too uneven or stony, but these tracts can be avoided, and may be cultivated in some other way afterwards. Wherever there is a spot that remains wet and soft, it should be drained by means of an open gutter, which can easily be filled up again if desired before the refilling of the pond takes place. Parts that cannot be reached by the plough should be turned over somehow, and the whole of the pond bottom should be harrowed as soon as the dryness of the soil will admit of it. Places should be marked that are by nature the best adapted for making beds, for the planting of the necessary future crop of vegetation. These beds should be prepared by carting on to them a few loads of suitable soil. This soil should not be taken from a tract of wet or boggy land, but from a dry corner somewhere. It should, if practicable, be of a different nature from that which forms the pond bottom, and the two may with advantage be amalgamated by digging. Where this soil is of poor quality it should be improved by some well-rotted stable manure, which may be dug or ploughed into it.

Such beds are excellent for water lilies and other strong growing plants. The whole of the pond bottom should have a final harrowing, and be sown with grass seeds. In many cases the grass will produce a paying crop, and should be cut when ready   In some, it may be cut two or three times with advantage, and when no longer available for this purpose, a flock of sheep may be turned on to it for a short time, to crop it close. In this way a fairly good sward will be produced.

When the pond is ready for refilling, which is best done in the spring, the bottom outlet should be closed, and, as the water rises, suitable aquatic vegetation should be planted at its margin. The object of this is to ensure the plants being submerged without loss of time. Aquatics rapidly shrivel, and become more or less injured if allowed to dry, or if exposed to wind or sunshine whilst out of the water. This may not necessarily destroy them, but care should be taken to avoid such occurrences. As the water continues to rise more planting should be done, and in this way a pond may be successfully stocked with a sufficient quantity of aquatic vegetation of the right kind.

The same soil does not suit all plants. Some will grow readily in a sandy or gravelly bottom, whereas others will not succeed at all under such conditions. There are many ponds which have sufficient good soil left in them, even after being cleaned out, to do without much further preparation in the way of carting in additional earth. In many cases, however, it is not so, · and this is one of the causes of failure I have met with. I have seen ponds with nothing but a mass of stones, many feet or even yards in depth, covering nearly the whole of the bottom, and in such ponds it is impossible to grow a sufficient crop of vegetation without first making proper beds for the reception of the plants. When the pond is nearly full the marginal plants should be put in, that is such as grow in very shallow water, or perhaps above water-mark occasionally.

Where new ponds have to be made their construction can be carefully studied, and the land that is to be presently covered with water used to the best advantage. Sometimes it is needful to make them by digging, sometimes by means of an embankment, or by a combination of the two. This must depend a great deal upon the nature of the ground, but in any case it is desirable to remove as little of the original sod as may be practicable. Should no digging out be required for the purpose of deepening or enlarging the pond, let the material for the embankment be obtained, as far as possible, anywhere but from the pond bottom. A moderate quantity of material may be taken from the deep end of it near the embankment, as a matter of convenience, if desired, but that is all. The object of this is to preserve the original soil and sod, so that on the filling of the pond it may remain undisturbed and form the bottom, and where this is done, and the water properly stocked at the right time by means of ova, fry, or yearlings, I dare prognosticate a most encouraging result.

There are many striking instances of the success attending the stocking of such waters, but the following will suffice. In May, 1884, 5,000 Loch Leven trout fry were put into Whitley Reservoir, near Wigan. The reservoir contained no fish and had been filled some time, and the water was consequently in excellent condition. On October 8th, 1886, the following appeared in the *Wigan Examiner* :—

"TROUT IN THE WHITLEY RESERVOIR."

" To the Editor of the *Wigan Examiner*.—Sir,—It may be interesting to some of your readers to know the particulars of an experiment to introduce Scotch trout into this water, and more especially so as several references to the matter have appeared in the papers. In the month of April,* 1884 . . . . I purchased 5,000 Loch Leven trout from the Solway Fishery, Dumfries, N.B. . . The trout reached Wigan in good condition, and . . . . were put into the reservoir immediately. . . . . Out of the 5,000 fish I have at different times taken, as specimens, six fish only. The first of these I caught early in the spring of this year, and it weighed thirteen ounces, while the last caught weighed two pounds six ounces†—truly a wonderful growth in so short a time. The six fish, when taken, had been feeding apparently entirely upon a species of shell fish which abounds in the pond. . . . .

" Yours truly,

" CHARLES APPLETON."

Along with the newspaper cutting I have quoted I received the following :—

* The fish were sent in May.   † Large numbers were afterwards taken, 2 lb. to 3 lb.

G

"Wigan, 14th October, 1886.

"J. J. Armistead, Esq.—Dear Sir,—I enclose you a copy of a letter which I have written about the trout I have bought from you, cut from the Wigan paper, which I think will be interesting to you. What strikes me as strange is the extraordinary rate at which the trout appear to have grown. And, from what I have been told, one has been taken out during the last few days weighing four pounds. Do you think it is possible such a large fish can have grown in the time from the fish which you supplied?

"I have exceedingly strong evidence that the pond contained no fish whatever before I stocked it, and yet it seems almost incredible that the fish now being caught can be part of those which I bought from you. I shall be much obliged if you will tell me whether, in your experience, such a rapid growth has occurred.

Yours faithfully,

"CHARLES APPLETON.

"P.S —There are millions of the shell fish mentioned in my letter in the pond, and all the trout seem to be gorged with them and nothing else.

"C.A."

This case does not by any means stand alone, although the growth is somewhat exceptional. It was equalled, if not exceeded, in the Dalbeattie reservoir at Buittle, in the county of Kirkcudbright. Trout fry introduced into this water in May were taken $1\frac{3}{4}$ lb. in weight a year the following September, that is in sixteen months, and later on 2 lb. The reservoir is small, and had the fish had more range they would probably have outweighed the Whitley fish in the same time. It had been newly made, and had, after filling, remained unstocked with fish for some time. On examination I found it to be full of freshwater shrimps (*Gammarus pulex*), on which the fish had been feeding. Had these reservoirs been properly stocked by means of eyed ova I believe the above weights would have been exceeded.

It is not at all difficult to obtain such results, and it can be done by so cultivating the water as to bring it into that peculiarly favourable condition for receiving the fish that mostly occurs where the surface of the ground has been left in its original state.

This highly encouraging condition of things will not last, however, unless due care be taken to cultivate those aquatic plants which are necessary to take the place of the submerged terrestrial vegetation when it has ceased to exist. This is a matter which is often seriously neglected, and which many people who have the care of fish ponds quite fail to understand.

The stocking in the Dalbeattie reservoir gave most encouraging results, and continued to do so because the proper conditions existed and continued to exist. In the case of a neighbouring piece of water known as Loch Fern, however, the fishing fell off considerably after the first few years. Now this lake was made by building a dam and flooding some fifteen acres of land, and the submerged vegetation continuing to exist for some time, a considerable amount of growth as well as decay of vegetable matter took place. The conditions were highly favourable for the generation of an enormous mass of fish food, but as the various plants died out these conditions ceased to exist. I gave instructions at the outset for the loch to be planted with suitable aquatic vegetation to take the place of the other plants, and had the instructions been carried out the result would have been very much better than was actually the case.

There existed on the estate close at hand a large crop of suitable plants, and I advised that a considerable quantity should be carted over and planted in the newly made loch. Instead of a horse and cart, however, a small hand basket was used, and the quantity transferred was quite inadequate to produce the necessary results. The fishing at first gave great promise, but instead of improving declined a good deal. The condition of some of the fish pointed to a lack of food, and as no improvement took place, but, if anything, rather the reverse, the water was run off, and a more barren wilderness could hardly be imagined than was presented to view. The plants introduced had made good progress, but occupied far too small an area of the bottom, and at my suggestion more were put in. The lake was refilled, and the plants have now been growing for some years. After the refilling the water was at first naturally very barren, but the condition steadily improved and the amount of natural food materially increased. I shall refer to this lake again in my chapter on stocking.

The advantages possessed by a newly-made pond where the sod has remained undisturbed are great, and if a good amount of aquatic vegetation were planted as the water was rising, such a pond should never show a backward tendency as regards the fish with which it is stocked. The water plants should be in sufficient quantity to take the place of the decaying land plants. In many cases it is not necessary to fill the pond to its highest level at first, and, where it can be conveniently arranged, the height of the water should be raised at the end of the second year. Early spring is the best time for doing this. Should the pond have been filled in September or October it is still better, even at the end of two years, to wait until March before raising the level of the water. The object in view in doing this is to submerge the grass above the original margin. It has been found that such a course is often followed by very gratifying results, and the reason is not far to seek. The opportunity should be taken for introducing a further quantity of aquatic vegetation, and some of the marginal plants may require lifting and moving further back. They will not have spread very much in two years, and, therefore, the work is soon done.

Since the necessity for vegetation in ponds has been made generally known, many owners of water, in their eagerness to have everything right, have introduced plants which, when allowed free scope, have increased to such an extent as to choke the water. In many cases this is very much due to a want of care in attending to them after planting. Many plants, if used at all, must be thinned out every year, from the second year, and sometimes from the first after introduction. This depends largely, of course, on the quantity planted in the first instance. But in any case such plants as *Potamogeton, Myriophyllum, Anacharis*, and others of a similar and free growing nature are better avoided altogether, unless it be intended that they shall be properly and carefully attended to. In their proper places these plants are valuable and are quite easily kept in order. But once neglect them, and let them get the upper hand, and a great amount of labour will have to be expended in order to bring them again within limits—and the labourers must understand their work, or it will, after all, probably be a failure.

We know how comparatively easy it is to keep a garden in order by attending to it at the right time, and how utterly hopeless the case becomes if weeds are allowed to go ahead and smother everything else. We also know how some of the worst weeds give little or no trouble if pulled out just at the right time. So with the bottom of a pond. The usual course taken in the case of a fish pond in the past has been to ignore the vegetation altogether, until it has become an absolute pest. Then some course has been taken which, owing to improper management, has tended to increase instead of to reduce the evil. I have seen weeds allowed to grow, flower, and ripen their seeds. Orders are given for them to be cut, and this being done, they are left floating till gradually they reach the shore on one side or the other, and are raked out. The seeds are thus scattered in all directions, and in a few years the pond contains such a mass of vegetable growth that the case gets hopeless. Now it will be apparent to my readers that such neglect can only be productive of unsatisfactory results.

My experience has been that in small ponds (say up to three or four acres) which are quite under control, some of the most free growing plants are highly beneficial. They require attention from time to time, and where this is given they are valuable helps to the welfare of a pond. Where this attention cannot be given, care should be taken only to plant such as do not require any cutting or thinning whatever. There are such plants, and they may be introduced with safety into any fish pond, as the utmost they can do is to cover the bottom with vegetation, thus producing a state of things that is much to be desired.

The plants chiefly to be avoided are those which put out fronds or branches, which rapidly reach the surface. In a large number of instances some one or other of these plants will sooner or later appear, and when this happens they should be removed. Fish ponds should, in fact, be weeded. In cases which have come under my notice a patch of vegetation has been pointed out which has grown of its own accord. More often than otherwise this has been a growth of pond weed *(Potamogeton)*. The edge of this weed bed has always been productive of a good fish or two, I am told. I say "Yes; but keep it in check and do not allow it to ripen any seeds." Sometimes my advice has been acted upon, and very good results have followed. In other cases the plant has been allowed to grow, and also to seed, and in a few years the case has been hopeless. Nothing short of a good clean out would do. Some of the plants of this class are very amenable to cutting, especially if the roots be disturbed at the same time. Care should be taken always to drag out the portions that have been cut. They naturally float and will be driven on to the lee shore of the pond by the wind.

There are several methods by which weeds may be cut in the ponds; if properly kept in check, which ought to be the case, there is no difficulty. A simple pruning hook on a long handle used from a boat will do good work in most cases. Sometimes the dragging of a weighted rope or chain over the bottom of a pond is quite sufficient, and I have successfully adopted this plan myself. It is not desirable in cases where it may do harm by dragging out plants that should not be interfered with. Where any of the free growing plants make their appearance naturally, as they sometimes will, it is, however, often very efficacious, and may be the means of preventing the eventual choking up of the pond. It is a good plan in some cases to drag the plants out by the roots, and for this purpose a specially constructed rake may be used. It may be made two feet wide and the teeth four inches long as a rule. Any blacksmith will make one, and the heavier it is the better; indeed, it is advisable to fasten a heavy iron bar on to it, to keep it well down as it is dragged behind a boat. It may have a bag-net attached, if desired, to pick up the detached weeds, and a more powerful apparatus may be worked from the shore, by means of a stout rope and blocks made fast to a tree or to a stake in the ground.

The advantage of the rake or drag is, that it can be used in places as may be desirable, whilst other parts of the pond may be left free. When properly handled it makes great havoc of the plants with which it comes in contact.

Water plants play a very important part in the economy of fish ponds. It is quite possible to maintain a large number of fish in a pond in which no plants are visible, but the difference between fish grown alongside of suitable aquatic vegetation, and those grown in water which is destitute of it, is marked. Where no plants at first sight appear to exist, there is often a considerable growth of minute vegetable forms, varying in degree according to the nature of the water and the general surroundings, but the presence of some of the larger plants is greatly conducive to the health of the fish. Livingstone Stone says :—"Water plants consume carbon and return oxygen. Trout consume oxygen and return carbon. By putting plants and fish together, therefore, we avail ourselves of one of Nature's great universal agencies, in balancing vital forces against each other, and maintaining the equilibrium on which the continuance of organic life depends." Speaking broadly, this is so, and it must be remembered that some plants produce more oxygen than others, and some afford food and shelter to a much larger quantity of animal organisms, and every trout culturist nowadays knows the value of *mollusca* and *crustaceans*, and especially some of the lower forms of the latter, including the *entomostraca*.

The fly-fisher in comparatively barren lakes, knows how often in the neighbourhood of patches of aquatic vegetation, the best fishing is to be obtained. The cause of this should be noted. Shelter, food, and a greater abundance of oxygen are temptations which cannot be resisted by the fish, and consequently they are sure to be found in proximity to the vegetation which produces them. As on land so in water, some plants are more favourable to the production of food than others, and these are the plants to seek out and cultivate, due care being taken to study their life history, and to avoid as far as possible those which might, by spreading unduly, become a pest in the pond. There is one plant (*Anacharis alsinastrum*) to which I have already just alluded, which has been found such a pest in some places where it cannot be kept under. It should never, on any account, be introduced into waters where due control over it cannot be maintained. In places where it can be kept down easily it is, however, a most useful plant. I have seen some lakes very much spoiled by it, becoming quite choked, and an excessive amount of decay taking place continually, and rendering the water very unwholesome. In these places it sometimes unaccountably dies out. I have seen a pond quite cleared of it in this way for awhile; but, a few small pieces retaining vitality will commence growing again, and spreading with rapidity, will soon choke the place up again.

Swans have been tried on some waters, and where they can reach it they make havoc amongst it. There are places in which it has appeared suddenly without any apparent reason. In such cases it has probably been carried by birds.

The *Anacharis* was originally introduced into this country from America, and is usually looked upon as the most free growing plant we have in our waters. There are some native plants that are as bad, if not worse, and one of these is the water milfoil *(Myriophyllum)*, which scatters its seeds profusely, and springs up, sometimes all over the bottom of a pond, and in a very short time spoils the fishing—indeed, I have seen it so densely matted together that it was with difficulty that a boat could be pulled through it. In such a case the simplest plan of dealing with it is to cut it down and drag it out, and where it becomes an excessive annoyance the best thing that can be done is to run the pond dry, and when the bottom is sufficiently free from water, to cultivate it roughly, and take a crop of grass off it for a season or two. The advantage of this course is very great, as the grass is not absolutely killed at once by the water when the pond is refilled, but continues to grow for some time. The conditions existing are very favourable for the production of a great mass of animal life, consisting of minute organisms, which are multiplied enormously for awhile, and the fish thrive on them.

In one way or another this experiment has now been often involuntarily repeated, and the result proved beyond a doubt, by the improvement of the fish themselves. Water plants are a great help in the growing of young fish, and in a pond well stocked with proper vegetation, trout fry will thrive where fish from the same batch of ova, kept in a bare pond and fed artificially, will sicken and die in large numbers. This fact alone may be taken as a fair test of the value of proper aquatic vegetation. As to the best plants to make use of, opinions may differ as yet, but one important point to be observed is the avoidance of those the introduction of which is likely to prove an annoyance rather than otherwise. About the margins of a stream, and at the point where it enters a pond or lake, experience has taught us the cultivation of the common water-cress *(Nasturtium officinale)* is highly beneficial. There are many terrestrial plants also, which

will grow to a considerable extent in water, such as white clover,
buttercup, etc. I found this out, as many other things have been
found out, quite by accident some years ago. The grass border
to a pond containing some trout fry had not been cut so carefully
as usual, and I noticed that where plants growing on the margin
of the water had put out their runners and were endeavouring to
grow in that element, there the little fish congregated and produced very much better specimens than were to be found in the
body of the pond. I have since several times cultivated these
land plants as marginal semi-aquatics, and with favourable results.
On another occasion, some trout fry which had accidentally
escaped from a rearing box were afterwards found in a water plant
bed, and in a month they had quite doubled the size of their
fellows, who were being reared by artificial means. The watercress was one of the plants which grew in this particular place,
and under the shelter of which the little fish had taken refuge.
Watercress requires to be planted where there is a comparatively
warm spring, in order to have a good crop early in the year, but
in many sheltered situations it will make good headway in April
by the heat of the sun alone. It can also be forced very rapidly
under glass, and it is worth while doing this about fry ponds and
their raceways.

Another excellent marginal plant is the marsh marigold
(*Cal.ha palustris*). Like the watercress, it may be planted at the
head of a pond, along the sides of the stream, and also all round
the sides of the pond itself, at any convenient places. It is a
useful plant for fry ponds, or any rearing ponds, and I have grown
it about my own for many years with advantage. Rabbits do not
eat it, which cannot be said of the watercress, for I have found
the latter quite exterminated by these rodents where the marsh
marigolds were left untouched. They are very easily cultivated,
and should be planted in spring, and they never become a pest,
which is very much in their favour. I have succeeded in getting
a variety that produces fine double flowers, and either these or the
single ones are exceedingly effective when cultivated in clumps.
They have the advantage of being very hardy, and once
established are easily propagated, either by means of seeds or
offshoots. I have seen trout fry harbouring under the shelter of

a tuft of this plant, in a manner which was exceedingly suggestive of its being a useful accessory to a fish pond.

The water plantain (*Alisma plantago*) is another marginal aquatic, which is very hardy, and invariably succeeds where properly introduced. It also grows freely in shallow water, say up to a foot deep, and, when planted at a depth of two or three feet, produces handsome floating leaves and excellent shelter for fish. Its natural position is at the margin of a pool or in shallow water, but it is well worthy of cultivation a little further into the pond. Should it grow too strongly it is very amenable to cutting, and is improved thereby as an aquatic; though for ornamental purposes, and as an insect producer, it is better to allow the plants full scope. It does not spread too much and can be readily thinned out in summer should such an operation at any time be considered necessary. The flower stems of this plant are rather handsome, and, if cut in September, will last all the winter through, and make handsome adornments for an entrance hall, or the corner of a room, or in smaller tufts, on the mantelpiece or sideboard.

CHAPTER III

Marginal plants—Insect life—Plants for deep water—Plants to avoid—Advantages of water lilies—Bottom-covering plants—A fish-eating plant—Ponds at Washington—Mollusca—Crustaceans—Eels—How to catch them.

M ANY of the marginal plants are very interesting, and all are more or less ornamental. The great spearwort (*Ranunculus lingua*) with its handsome yellow flowers is one of these.

The lesser spearwort (*Ranunculus flammula*) is another plant which grows freely at the edges of lakes, and although its natural habitat is oftener in bogs and ditches, yet it is a very useful little plant, which I have frequently grown extensively in raceways and fry nurseries. When submerged in water a foot deep it sends up long stems supporting floating leaves and becomes an interesting looking aquatic. I have often recommended it for the aquarium, where it has given great satisfaction. It is very hardy, and will grow in almost any place where there is moisture. There are several closely allied species.

Another plant that is easily grown is the brooklime (*Veronica beccabunga*). This also is a very useful plant, and, being hardy, succeeds anywhere. It is said to be very good for cattle, and it is certainly in its right place at the edge of a fish pond.

One of the most beautiful of our marginal plants is the bog bean (*Menyanthes trifoliata*), which grows either in the water or out of it. It delights in the border of a lake where there is a bed of soil or turf from which it can spring, and from this it will grow out as a floating plant, sometimes for many feet. It is very amenable to cultivation, and the exquisite flowers are often much improved thereby. Before opening they are of a bright rose tint, changing to a most delicate pinky white when expanded. The stem or root of this plant possesses medicinal properties, and makes an excellent tonic.

The iris family is a very interesting group of plants, and one which produces numerous varieties, some of which have exceedingly handsome flowers. Our native species, the yellow iris (*Iris pseud-acorus*), is one of these, and does well when planted in masses on the banks of a lake. It will grow in water up to a foot or two feet in depth, and is a desirable plant to have about fish ponds. It succeeds also very well growing from the side of a perpendicular bank bounded by deep water, and in such a situation produces an excellent shelter for fish, which like to frequent its mass of semi-floating roots. I have known perch to deposit their spawn upon them. We are indebted to many foreign countries for contributions to this family, and when grown as ornamental plants some of these are very handsome, and make most interesting groups with a rich variety of colouring that it would be difficult to excel. They ought to be much more extensively cultivated than they are, and anywhere near the house are well worthy of the attention of the gardener.

The bullrush (*Typha latifolia*), as it is commonly called, is an exceedingly handsome plant, which like the iris, will grow in water up to two feet or more in depth, and is to be recommended on account of the shelter which it affords, and also as a food producer. Where ornaments is an object a few good clumps of this plant are most effective. The lesser bullrush (*Typha angustifolia*) is a smaller plant, and is much more uncommon than *Typha latifolia*. Where allowed full scope it is perhaps the most desirable of the two, and is much more graceful in appearance, its slender stems and smaller heads possessing a peculiar beauty which is not so apparent in the other. Both are equally useful in their proper places, and are easily kept within bounds.

The flowering rush (*Butomus umbellatus*), now unfortunately becoming rare in this country, is well worthy of the attention of the aquaculturist, and its handsome flowers soon make it a favourite when it becomes known. It is very easily grown, and thrives best where well rooted in a good bed of soil.

There are many other plants which do well in similar

situations to the foregoing, but those I have mentioned have been tried, and their introduction has proved beneficial. The subject of pond culture has been so little studied, that the uses of many plants are not yet thoroughly understood, and care is needed in the introduction of new and untried species. Whilst treating of marginal species, I would point out the advantage arising from the planting of suitable insect producing plants, shrubs, and even trees. Where tree planting is done in the vicinity of a lake or pond, it is very desirable to put in a few suitable specimens near the water. The oak produces a great amount of insect life, and a few alders are in their element by the pond side, and do good. Osiers and willows are also excellent, and, if the right kinds be planted, they may be cut annually, and in this case prove remunerative.

It is well worth while to bestow a little care on the vegetation surrounding lakes and ponds, for a large amount of insect life may be produced by encouraging the growth of suitable plants. In the case of willows, shrubs, and tall growing plants, it will be found a great advantage to keep them back a little way from the water, so as to leave plenty of room for a path along the edge of the lake. The leaves will not then fall into the water so much, but thousands of insects will, and these will be very good for the fish.

The introduction of the plants that inhabit comparatively deep water requires much more consideration than is necessary with marginal specimens. The latter can easily be got at, and there is no difficulty in keeping them in order. They are more constantly seen, growing as they do partly above the surface of the water. We know what weeds are in a garden, and how freely they grow. They often grow just as freely at the bottom of the water, and require attention just the same. There is one difference, however, and that is, that when a pond bottom has been properly prepared before filling, the weeds are often many years in making their appearance. When they do commence growing they usually appear in one, or at most, two or three places only, and are at first very easily dealt with. The plants most suitable for introduction, therefore, are those which do not spread too rapidly, and which in spreading do not choke the water to the injury of the fish.

It is quite easy to find such examples, and one which I have used most successfully is the stonewort *(Chara flexilis).* There are several closely allied species, and I merely take *C. flexilis* as an example. It soon covers the bottom, where it grows, and as it does not usually attain a height of more than two feet or so, it cannot be in the way of the fly-fisher. Even when trolling it is not in the way, although it may occasionally be brought up on the hooks should they sink too deep; but even then there is advantage rather than otherwise, as the plant does good service in keeping the hooks off the actual bottom. It also provides good shelter for those creatures on which the trout live, and I have often seen it much frequented by *Limneæ*, etc. It grows well in either standing water or in a stream. Sometimes it suddenly disappears from a pond, but in such a case some will be found in the stream below, which will soon raise another stock, if removed and planted near the head of the water.

We often find lake bottoms covered in places by some member or other of this family *(Characeæ),* and there are other plants which are also often found, naturally, in similar situations, such as *Myriophyllum* (milfoil) *Potamogeton* (pond weed), etc. These, and other free growing plants, I have already alluded to. They should, as a rule, be avoided, although in large lakes where they have plenty of room, they often do very good service. I have seen ponds entirely choked by *Potamogeton crispum* and *Potamogeton natans,* and yet both are useful, especially the former, in their proper places. The difficulty is to keep them within reasonable bounds. For coarse fish ponds I have found *Potamogeton crispum* very useful, but it requires a good deal of attention.

Of the taller growing plants, the white water lily *(Nymphæa alba)* is one of the best. If properly planted it grows readily, and the floating leaves provide an excellent shade for the fish, under which they delight to swim, and where they find also a quantity of natural food. The beauty of the flowers alone is very much in their favour, while they spread very slowly, and should the floating leaves at any time be in the way, it is a very easy matter to cut them. They will thrive wherever there is a fair depth of good soil, but do not grow well on a stony or sandy bottom—

indeed this applies to most plants. The native species *(N. alba)* does best in Britain, but some of the introduced kinds will do very well in some waters, and amongst these are the sweet-scented water lily *(N. odorata)* and the Cape water lily *(Aponogeton distachyon)*, which is a little gem.

There is a British plant which is very little known, being only found in a few localities, and which is both very pretty and useful in ponds. This is the fringed water lily *(Villarsia nymphæoides)*, the leaves being small and varied in colour, from green to reddish brown, and the flowers yellow, and about the size of the common yellow cistus. It is very easily cultivated, and spreads more rapidly than the white lily *(N. alba)*.

The yellow water lily *(Nuphar lutea)* is a freer growing plant than *N. alba*, and the leaves are larger and the flowers not so showy as those of its congener. It is, however, a very useful plant, and is easily cultivated where there is a good bed of soil into which it can strike its roots.

Water lilies are easily planted in artificial ponds or lakes whilst the water is rising, but it sometimes becomes needful to introduce them into places which cannot be run dry. It is a very good plan in such cases to plant them in very light wicker baskets, and sink to the bottom of the lake. The basket will rot, but the lily will grow, and soon takes hold of the bottom. The simple tying of a stone to the roots is, however, sufficient to sink them, and they may then be gently pressed into the soil at the bottom by means of a pole, care being taken that the crown of the plant is uppermost.

There are some plants which nature seems to have specially provided for the covering of pond or lake bottoms, and first among these is the lakewort *(Littorella lacustris)*. This little plant will never become an annoyance anywhere. It only averages about three inches in height, and is very rarely found to exceed six inches—indeed, it only attains to this height when growing in peculiarly favourable places. I have found it occasionally reaching the latter dimensions, but only in some corner where the temperature and soil have both been specially favourable, and such instances are unusual. It is one of the most useful little plants that I know of in a lake or pond, and soon covers the bottom with a rich grass-like carpet, and at once forms a refuge for a large quantity of animalculæ, shell fish, etc., the latter attaching their spawn to its leaves very freely. It grows readily in almost any reasonable depth of water, say from a foot to ten feet or more, and is one of the most harmless plants we possess. It can never interfere with fishing in any way, it never needs thinning, and no matter how much it may spread, it can never become a pest. Being a valuable food producer, it is an exceedingly desirable addition to our fish ponds, and should be largely planted. One of the best methods of doing this is to lower the water a little, and dibble in the plants in small groups, varying from three or four to twenty plants, according to the quantity to be planted and the size of the pond The plants spread laterally, so that it is only a question of time for them to cover the whole bottom of the pond. In waters which cannot be lowered, the plants may be tied to small stones, and dropped into the water in suitable places. When this plan is adopted, several may be tied to one stone, care being taken that they are not bound too tightly. They may also be planted in rough shallow boxes and sunk in the lake, or in flat baskets, which may be treated in the same way. I have known a case in which a large number were sorted into bunches of eight or ten, and these consecutively tied on a string, somewhat like the papers on the tail of a kite. A stone was tied to each end of the string, and it was then sunk in the water, care being taken to stretch it well, so that the plants could not rise, but must stay on the bottom, into which they soon sent their roots.

The next plant which is to be recommended is the charming little water lobelia (*Lobelia Dortmanna*), and the only objection that can be raised to it is that it sends up a single flower stem above the surface of the water. This drawback, however, is not worth a moment's consideration, as they can be readily cut, and will not again reappear the same year. With the exception of the flower stem it only reaches a height of some two inches, and does not spread unduly. It is indeed a very desirable little plant to introduce, and is propagated by seed. The root consists of numerous white fibres, and the leaves, which are fairly numerous are arranged in a somewhat star-like form at the bottom of the water. Shell fish deposit their spawn freely upon its leaves. Of all plants, the lakewort and the water lobelia are, I think, two of the most suitable for introduction into trout ponds. The latter was named in honour of the Flemish physician, Matthias de Lobel, who was botanist to James I.

The great water moss (*Fontinalis antipyretica*) is a valuable plant, and delights in rocky streams and raceways. It also does well in ponds where there is a good current, and grows chiefly on stones, though it may often be found in streams growing on wood or tree roots, and I have found it growing freely on gravel beds in lakes. It is very useful in providing a good shelter for trout, and also is an insect producer, and in raceways is invaluable. One great advantage is that it does not grow with sufficient luxuriance to choke places up as some plants do, and it is easily thinned out if desirable. It can be introduced by planting stones on which specimens are growing, or by placing tufts of it in the gravel. Planting in raceways and streams is easily done by simply fastening a stone on to the root of the plant and pressing it into the bottom.

In addition to the great water moss, the starwort (*Callitriche*) and water crowfoot (*Ranunculus aquatilis*) are useful in streams. The latter sometimes becomes a pest when it gets into ponds. It is needful, therefore, to use care in introducing it into pond or lake tributaries, and the same applies to some other plants. The milfoil (*Myriophyllum*), for instance, is an excellent plant in many streams, but in ponds it is a perfect pest. When it occurs in a stream which feeds a pond, the seeds or small portions of the plant itself will inevitably float down into the pond and grow. Some of the *Ranunculaceæ* are exceedingly acrid and have been used for producing blisters, but the wounds made proved very troublesome and difficult to heal. *Ranunculus flammula* already referred to is one of these, though in trout ponds it does not seem to be injurious, but rather the reverse. The water celery or celery-leaved crowfoot (*R. sceleratus*) I have carefully avoided, as well as some others of the genus, on account of their acrid tendencies.

The bladderwort (*Utricularia vulgaris*) has, been found to eat fish, or at least to destroy them. This discovery was made by

H

Mr. G. E. Simms, of Oxford, who had the plant growing in an aquarium in which the fry of perch and roach were also kept. Seeing some little fish apparently lying dead amongst the bladderwort, Mr. Simms endeavoured to remove them, and found that they were firmly held by the plant. The latter may be briefly described as a floating species growing in ditches, ponds, and other still waters. The stem lies in a somewhat prostrate position, from which numerous thread-like leaves grow out in whorls. Amongst these leaves are found a number of tiny vesicles, which were formerly supposed to be air-bladders by means of which the plant floated to the surface. These now come out in their true character as traps for living creatures, including fish, upon which the plant probably feeds. The vesicles are somewhat pear-shaped, and possess an opening at the smaller end which is closed by a valve. This valve opens on pressure from without, and closes upon any unfortunate creature that may enter. Young fish have been found caught by the head, tail, and umbilical sac. *Crustaceans, larvæ,* and other organisms, are entrapped by these vesicles, or so-called bladders, from which the plant derives its name. In summer it sends up a flower stem some five or six inches high. The blossoms are yellow streaked with purple.

Mr. Simms tells me that experiments have shewn that it is very seldom, indeed, that young roach become entangled in the bifid processes with which the quadrangular mouth of the vesicle is furnished; young perch, however, succumb freely to its influence, which is a purely mechanical one. Under a high power microscope, the vesicles appear perfectly smooth and polished, and they seem to be armed with a series of reversed serrations, pointing towards the opening. These serrations catch the delicate skin of the fry, whose every struggle only sends them further on towards the opening, and so on until the posterior wall of the vesicle is reached, when the fish is prevented from making its exit by the closing of the trap. Sooner or later the fish dies, decomposition sets in, the tail drops off, that is, as much of it as is exposed outside the vesicle, whilst the portion that remains within is largely utilised for the future support of the plant. Owing to its having no root, it is dependent to a considerable extent on the amount of nutrition that it can obtain through the medium of its vesicles, which form a series of little stomachs, both in shape and function, along the branches in the axils of the leaves.   The walls of the bladders are almost opaque, save at the extreme of the posterior end, which is semi-transparent, and forms a spot of greenish light.   This undoubtedly arouses the curiosity of the little perch, which in their anxiety to institute further investigations leads them to become entangled on the bifid processes, where all further chance of their ever troubling an angler is abruptly ended.   Care must be exercised in collecting the plant for experimental purposes, as a short exposure to the air seems to cause the traps of the vesicles to lose their elasticity, and renders it valueless for killing purposes.

The thanks of fish culturists are due to Mr. Simms for making known the result of his interesting work, though, as far as trout culture is concerned, there is probably little fear from the presence of bladderwort, beyond the fact that it consumes a certain quantity of trout food.   In ponds used for coarse fish culture, however, the case may be very different, and from these it should be carefully excluded.   In the United States, as soon as the character of the plant was made known, the late Professor Baird issued a circular to American carp culturists warning them to remove it from their waters.   In ponds such as those of the United States Fish Commission at Washington, which I had the pleasure of visiting in 1893, such a plant is calculated to do harm, for out of the multitudes of small fry which crowd some of them, a considerable number would probably fall victims to its killing propensities.

In the ponds at Washington water-plants are largely cultivated, and are a most needful part of the system.   The ponds are carefully surrounded by fine wire netting to keep out the rats, tortoises, water snakes, etc., and are well watched and tended. They have proved a most successful part of the work of the United States Fish Commission.   In some of them the plants on which the fish deposit their ova are grown in trays, and when the operations are over these trays are lifted out and placed in other ponds, where the little fish find plenty of food, and are not devoured by their parents.

When once a piece of water contains a proper assortment and amount of plant life, there will be a good chance of maintaining a sufficient stock of those creatures which form the natural food of trout. *Crustacea, mollusca,* and other forms, should be plentifully present in the water. The two former are very easily introduced, and should be found in every trout pond. They are very prolific, and when once put in soon multiply and stock the water. The result of having a good supply of these creatures has already been referred to in the case of the Whitley and Dalbeattie reservoirs, which are by no means exceptional instances of the rapid development of trout. The famous gillaroo trout, named by some naturalists *Salmo stoma-chicus,* on account of the toughness or thickness of the middle coat of its stomach, has been found to feed very largely on shell fish. The late Dr. Francis Day says :—" Its stomachs are occasionally served up as gizzards. Thompson obtained from the stomach of one example, about eight inches long, about a thousand shells of *Limnea peregra, Valvata piscinalis,* and a few specimens of *Sphærium corneum.*" The well-known pink colour of the flesh is undoubtedly attributable to the food on which the trout live, and it has been found that the way to produce both a pink colour and a delicate flavour is to feed largely on *mollusca* and *crustaceans.* I once took in hand by way of an experiment the feeding of a pond full of American trout (*Salmo fontinalis*) entirely on shrimps from the sea. These creatures were obtained by bushels and fed to the fish. The diet was an expensive one, but the effect was most satisfactory, resulting at the end of six months in beautifully pink-fleshed and delicately-flavoured fish. Strange to say, the trout took a violent dislike to these shrimps when unboiled, but boiled ones they eat eagerly. A diet of shell fish also produces very good results as regards flavour and colour, and the rate of growth is far above the average.

Shell fish (*Mollusca*) may be introduced into any pond or suitable stream without the slightest difficulty. They should be put into the water in various places, where there is some vegetation on which they can feed. They deposit their spawn freely on water plants, any pieces of rotten wood, or even on stones, and when once a brood has been secured they are never likely again to be eradicated. Some kinds do much better than others, and among these perhaps one of the best is the *Limnea peregra.*

Crustaceans in the shape of fresh water shrimps (*Gammarus pulex*) are also very easily introduced, and should be put into the stream flowing into the pond, where they will at first be more at home, and they or their progeny will soon drop down into the pond and stock it. The fact must not be overlooked that some *crustaceans* are parasitic on fish. Some knowledge is therefore requisite, before introducing living creatures into a trout pond, or a serious mistake may possibly be made. I once had a consignment of fish from abroad, and on arrival they were closely examined before being turned out, and were found to be infested by a leech. These creatures caused the fish much annoyance, if they were not otherwise hurtful. Each fish was carefully examined and the pests removed, and after being passed twice through salt baths, and performing quarantine, they were found to be quite clean, and were then turned out. Seven of these leeches were put into a glass jar for further examination, and a couple of sticklebacks were introduced. A leech immediately stuck to one of them, having hold of the jar at the same time, and in spite of the utmost endeavours of the fish, the leech held it for fifteen seconds or more, and then relinquished its hold of the glass and stuck to the fish with both extremities, to its evident great annoyance. The others soon followed suit and took hold also. Next day, however, the tables were turned, four of them having taken inside berths. The remaining three still adhered firmly to one fish. These leeches, when examined under the microscope, were found to be infested by scores of smaller parasites.

> " Great fleas have little fleas
> To live upon and bite 'em ;
> Little fleas have lesser fleas,
> And so *ad infinitum.*"

This occurrence suggests the exercise of great care in importing foreign fish lest we import foreign parasites and plagues, as so frequently in the past. All imported fish should be carefully examined, dipped, and quarantined, before being passed into British waters.

In the cultivation of fish ponds eels must not be overlooked. They will manage to get into some ponds, and often do grievous damage. I once turned 1,760 yearling trout into a pond in May, and in August they were taken out again and only 1,220 were forthcoming. A couple of eels, which weighed 3lb., had accounted for over 500 of them. This is but an example of the enormous amount of damage these fish will do if left to themselves. A correspondent wrote me quite recently that he had lost eighty per cent. of his yearlings, owing to some eels having found their way into the pond. Ponds should, therefore, be made secure against these fish, and where this cannot be done, or where their presence is even only suspected, eel traps should be set. Those I have successfully used consist of wicker baskets, sunk in the water in suitable places, and buoyed to assist in finding them again. Where a number are fixed, half-a-dozen or a dozen may he strung on a line a few yards apart. These require baiting, and this may be done by placing in them some pieces of boiled horse-beef or fresh rabbits' paunches. If herrings are procurable, half a herring or a whole one in each trap is a killing bait. It is not really necessary to sacrifice the herring ; the refuse, after they are cleaned for cooking, answers equally well, if not better. It is a good plan to place the bait inside a tin cannister punched full of holes, but the trap, and especially about the entrance, should be smeared with herring also. In some places worms make an excellent bait, and the way to use them is as follows :—Get a lot of worms and some small sods, and place them sandwich fashion in a box for an hour or more. The worms will enter the sods, which then make excellent baits. I have had traps worked in this way, and they have caught enormous quantities of eels, sometimes getting the traps packed full. I have also heard of strings of worms being tied into bunches, but never tried the plan. The warm months from April to October are those in which the traps work well. During the cold or winter months they do not act at all in some localities.

CHAPTER IV.

*Preparation—Stocking—Carrying live trout—Dipping the trout—Transit—
Large fish—Two-year-olds—Yearlings—Fry—Nursery ponds—Water plants—
Turning out fry — Fry in rivers — Excellent travellers — Glass carriers —
Advantages of—Equalising temperature—Fish killed by thoughtlessness— Wooden
carriers—Metal—Travelling trout in August—Care required—Fully eyed ova—
Trout at the Antipodes—American work—Successes.*

ANYONE who has any knowledge of farming or gardening, knows perfectly well that unless the ground is properly prepared, and the seed sown at the right time and in the right way, the future crop will probably not be a great success. Just so it is with fish ponds.

Attention has already been drawn to the needful preparation of the ponds, and this having been duly attended to, the water is ready to receive the crop. It may be produced from eggs (ova), or by transplanting the fish themselves. The latter course is the one that has hitherto been most generally adopted. There are cases in which owners of ponds do not wish to wait, but to have a crop of large fish at once. In such a case winter is the best time for the fish to be transferred, and if possible, during a frost. The fish are then more easily carried from place to place, and will bear handling better.

Years ago an idea prevailed that in carrying living trout, the two things needful were to aërate the water vigorously, and to change it whenever possible. Thus we find many elaborate contrivances have been invented for this purpose. For all the trout I ever carried—and I have travelled with them to all parts of Britain, long journeys, both by sea land—I never used any aërating machine, but an ordinary lading-can of about a gallon capacity. I believe a pair of bellows may at times be useful, but I have never yet required them. The great secrets of carrying fish successfully are attention to temperature and an empty stomach

To those who understand the condition of a fish, and its relation to the water under such circumstances, the reason is obvious. I have conveyed hundreds of trout up to 3 lb. weight without loss. Before such large fish are turned out on arriving at their destination, they should be dipped in a saline solution. Permanganate of potash may be used with success for this purpose, and chloride of sodium or common salt does very well. In cases which occur, where the sea is not far distant, sea water is very good, and it answers the purpose admirably.

I have dipped many thousands, and am quite satisfied as to the great advantage often arising from such a course. The farmer finds it pay to dip his sheep, and the fish culturist his fish: indeed the latter may learn a great deal from the former, if he will only keep his eyes open. When fish, on arrival at the pond side, require to be dipped, a very good *modus operandi* is as follows :— [Of course it is understood that I am now alluding to large fish only.] A tub should be in readiness at some suitable place on the bank of the pond or lake, and this tub should be filled with a solution of salt (1 lb. of salt to fifteen gallons of water is a good mixture). The fish should then be placed in the tub by means of a landing-net or by hand, and allowed to remain there until they begin to look sickly, or turn over on their sides. They should then be taken out, and dexterously pitched into the pond, where they ought to recover themselves at once. Sometimes after a long journey, the fish are a little sickly to commence with. When such is the case, they should be placed in some sort of a cage in a stream for a while, so as to get refreshed before going through the salt bath. A good-sized hamper will do very well for this purpose should nothing better be available, and it should be placed where a good current will flow through it. The apparatus which I prefer for the purpose, however, is a good large landing net, and the one I use and in which thousands of fish have been treated, is rectangular in shape. The frame, which is of iron, measures five feet by one foot eight inches, and the net bags about fifteen inches. A dozen, or a score of fish can lie in this net comfortably, and the handle is placed on the bank of the pond, with a stone on it to keep it in position. Where the bank is too steep to admit of this being done, the handle may be placed over the water, or stuck into the bank, and the whole supported so that the rim of the net is a little above the surface. The fish will soon recover on being placed in such a receptacle, and may then be safely dipped and turned into the pond.

The only case in which I remember being in any difficulty with large fish in transit was once during some hot weather early in October, when I travelled a lot by special request, but in opposition to my own advice. I lost eight on this occasion out of two hundred.

Two-year-olds are good fish for stocking purposes, but they require a large bulk of water, and the risk attending their transit is greater than in the case of yearlings, which can now be travelled with perfect safety, and quite easily to any part of this country, or on to the continent of Europe. Yearlings are undoubtedly the best fish for stocking purposes, inasmuch as they may really be called fish, which can perhaps hardly be said, from a practical point of view, of fry. It is in the fry stage that the great loss occurs, which all fish culturists have to guard against by every means in their power.

A difficulty that has for years baffled the attempts of both scientific and practical men to bridge over, is no small one, and although it has now been overcome by experienced fish culturists, yet it still remains a source of considerable danger. A few hours' neglect, or a little carelessness or mismanagement, may sacrifice ninety per cent. of the little fish, and it will thus be seen that, in the hands of the inexperienced, trout in their infancy stand a pretty good chance of being killed, often, it may perhaps be, with too much intended kindness. Yearlings are as safe as fish can be, and unless very clumsily managed by the recipient, there should be little or no risk attending their introduction into new water.

It is well known that trout are keenly sensible to temperature, and that a very rapid change, such as a sudden transference to water a good many degrees warmer or colder, is very prejudicial, and will sometimes even kill them outright, and that very speedily. They are frequently travelled in iced water, so that the chances are that the water into which they are about to be introduced is higher in temperature. In certain cases ice is necessary to ensure their safety during transit. On arrival at their destination, it is always safe, although it may not be always necessary, to fill up the cans containing the fish with water from the river or lake they are to inhabit. Then pour off about a third of it, and fill up again, and if the difference be known to be great, repeat the process a second and even a third time. The fish may then be safely turned out.

Yearlings are a very good size or age with which to stock water, and over ninety-eight per cent. of the consignments from the Solway Fishery reach their destinations in safety, although sent quite alone, and to all sorts of places, from Land's End to John O'Groats. In Autumn and winter they travel perfectly. In spring, as the days get warmer, there is risk, and during April and the early part of May, when yearlings are sometimes travelled, very much depends upon the state of the weather and atmospheric conditions. The influence of the sun on the water is often considerable, and after a week of warm days and nights at this time of year the fish will be feeding freely and growing also. They have to be tanked and starved, and the temperature of the water lowered some fifteen degrees by means of ice. This ordeal is trying to them, and when at last they are sent on their journey, perhaps amidst thunder and lightning, they must run some risk. In this favoured country we have cold nights in April and May, and during such, yearlings will bear transit until late in the season ; but once let the weather become thundery and the danger is largely increased, as also is the cost of transit.

There is no real difficuly now-a-days in the mere work of conveying fish, even in the month of July, if placed in charge of experts. It is simply a matter of cost, and the expenses at that time of year would, I need hardly say, be considerable. Two-year-olds, if the way be tedious or complicated, are not absolutely safe without an attendant, for part of the journey at least. Should they become sickly *en route*, a little attention soon restores them, whereas if left unattended many of them under such circumstances would succumb. Yearlings are found in practice to give excellent results, although I have many cases in which fry have turned out equally well. I say equally well ; perhaps I ought to have said better, for although the loss by death has been numerically greater, yet, comparing costs with results, they have compared favourably with yearlings. Strong, well bred, and healthy fry, are very good for stocking waters that do not contain any fish.

In cases where other fish are present it is needful to take care that the fry are not devoured by them, and this may be done by planting them in the stream feeding the pond, or in some small tributaries, where they will soon learn to take care of themselves. Instinctively they soon do this, and I will guarantee that a lot of fry in good condition will, if properly turned out, make considerable addition to the crop of large fish in the future. It is because fry have been injudiciously turned out, on the one hand, that they have often not produced the desired results, while on the other they have not been in condition, and so have had little chance, if any, of surviving all the dangers to which they are exposed. I have again and again met with people who have hatched a quantity of trout ova, and kept the fry in rearing boxes, where they have been fed until they began to die off. They have then been at once turned out into the lake or stream for which they were destined. Is it to be supposed for one moment that they would cease dying after being turned out? I say, no! most emphatically. Probably it may be the means of saving some of them, but the mischief has been done, and it is impossible to undo it. Fry so reared are not in condition for turning out, and therefore the planting of them results in failure. They have been fed on various artificial foods, and the working of their delicate little stomachs sadly disorganized. This course of unnatural diet has at last upset them to such an extent that they begin to die off. It is much better to turn them out just before they begin to feed, that is, before they have quite absorbed the umbilical sac.

Stocking with yearlings is a simple and easy matter; any intelligent keeper can do the work successfully. But with fry it is an entirely different affair. It does not matter how intelligent the operator may be, unless he has a fair knowledge of what he is doing, the chances are that some mistake may be made which will frustrate his good intentions. I have now stocked the same waters with fry for many years in succession, and have got excellent results, but they have been introduced at the right time, as well as in the right place, and in the right manner.

In the case of a lake or river in which it is intended to keep up the stock of trout by a periodical planting of fry, it is of importance to have nurseries prepared for them, and these require to be properly made, or they may be worse than useless. Their construction is often a very easy matter. A series of long narrow ponds of the simplest description will do, and a couple of good men should make enough of these in a few days to stock a lake of twenty or thirty acres. In the case of large lakes it is desirable to have a larger series of these nurseries, and sometimes, in that of more extensive waters, it is an advantage to have them at several places. Once made they require little work to keep them up, and can be used year after year with excellent results. Some lakes are fed by streams flowing a considerable distance through flat marshes or meadows. In such a case it is desirable to go further away, until a place is reached where the water has a slight fall. It matters little about distance—it may be a hundred yards, or it may be half-a-mile. In the latter case the little fish will easily find their way down stream. When the sight has been chosen, make a long raceway or aqueduct with alternate stretches of deep and shallow water. It does not matter about the wide and narrow stretches being symmetrical at all ; they will have to be made to suit the level of the ground and surrounding circumstances. It may be necessary to have them longer or shorter, but where practicable, from forty to sixty feet will be found a convenient length for the wider and deeper parts, and if carried out somewhat according to the accompanying plan the raceways will be a little longer of the two.

Let AA represent the stream supplying the pond E. At the point B cut a raceway eighteen inches wide and a foot deep. Dig out some ponds CCC two and a half feet deep by three and a half feet wide, and connect them by raceways DD, of the same width and depth as B. Where sufficient fall can be obtained, each of these ponds or nurseries should have a bottom outlet at the most convenient point, so that they can be run dry at any time if desired. A four-inch pipe answers best for this. A screen should be placed at the point B to prevent larger fish coming down and another between D and E, which will allow fry to pass down into the lake E, but will prevent larger fish getting up into the ponds. Keep the bottom of each pond nearly level for two-thirds of the length, and let the tail end slope upwards to the level of the raceway (one foot deep.) This slope may be covered with gravel. About the centre of each pond place a four-inch pipe or tile lengthways on the bottom. I use four-inch sanitary pipes. If these be selected, place the pipe with the flange up-stream, and over it build an embankment of sods up

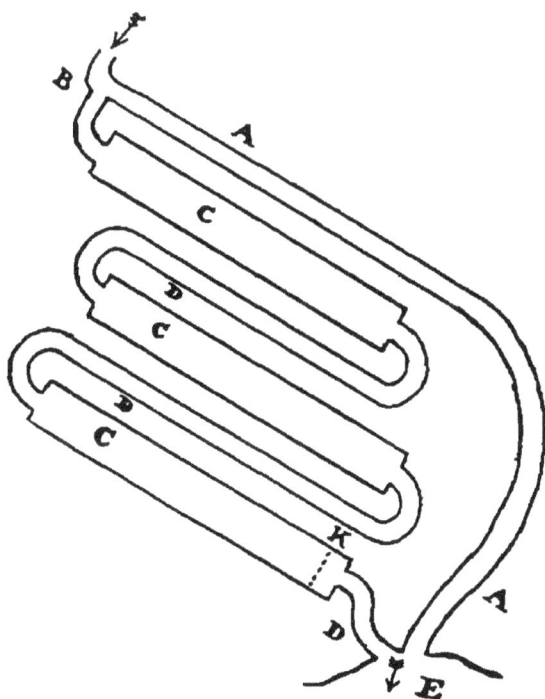

Fig. 3.

to the water level, leaving a gap in the middle. Take care to make it tight, so that the water can only pass it by flowing either through the pipe or over the gap in the sod bank. The sides of the ponds may be perpendicular, or bevelled, as may be most convenient. Where the nature of the soil will admit of it, make them nearly perpendicular; should it be necessary to build them, do it to a large ex ent at least with sods.

The sides of the raceways should be planted the whole length with suitable vegetation. Watercress (*Nasturtium officinale*) is excellent for the purpose, and also the marsh marigold (*Caltha palustris*). The pretty little golden saxifrage (*Chrysosplenium oppositifolium*) may also be used with advantage, and the brooklime (*Veronica beccabunga*). There are many land plants that do very well in water for a time, and I have found that the decay of these plants is favourable to the growth of animal life (see page 88). No floods or freshets should be allowed to sweep through these nursery ponds. This can easily be guarded against by a sluice, or what is more simple, allowing the water to enter them through a four-inch pipe.

They should be made as tempting for the little fish as possible, by planting aquatic vegetation about them. It may sometimes be desirable to place a screen across the end of the lowest pond in the position of the dotted line at K, but should the fish be found to collect above this screen they must be fed or they will die, and in feeding them all pains should be taken to coax them away from the screen and get them up to the head of the pond. As soon as the bulk of the fish have settled in the ponds, remove the screen and let them have full scope to go where they like. These ponds may be made accessory to an artificial ova bed or a spawning race, or they may be stocked with fry that have been purchased or hatched elsewhere.

The screen, if used, should be made loose, to slide in a grooved frame, or it may be simply fixed by embedding the frame in the sides and bottom of the pond. It will only be required for a short time after the fry are introduced, and may then be taken out and stowed away to use another season. It will be seen by the diagram (Fig. 3) that the ponds and raceways are arranged in the most compact form, instead of having them ranging over a long expanse of ground. This is an advantage, as they are readily inspected, and also can be more easily protected from the depredations of herons, kingfishers, etc. A few covers made of rough boards or basket-work may be advantageously used, and a few pieces of wire netting should at any rate be placed over the most exposed portions of the raceways. A little care of this kind will be well repaid. The covers not only serve as protection but as shades for the fish, and it is desirable to allow the water to flow over some of the embankments of the ponds while it passes underneath others. This is easily managed by having a few rough plugs well fastened to the end of a stout stick or piece of rail. A piece three inches by one inch will be found to answer well, and the wooden plug (Fig. 4) should be made so as to fit loosely into the flange of the pipe.

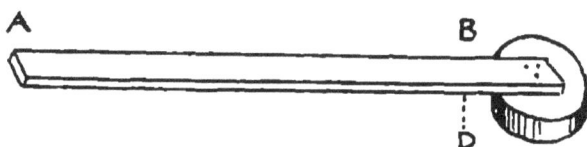

Fig. 4.

The handle A B acts as a lever in drawing the plug, and should be three feet long. When required place a similar piece of wood crosswise below the handle at the point D to act as a fulcrum, and holding it firmly, or even pushing it forward a little while the end of the piece A is drawn back, the plug comes out of the socket of the pipe, and can then be lifted out. This answers well also for working the bottom outlets of the ponds.

When the ponds have been prepared and the screens adjusted, and they are all in order for receiving young fish, the latter should be gently introduced at the head of the first pond in the nursery, and left to themselves. They will probably remain there for a time, but will soon scatter, and many of them will drop down from pond to raceway, and from raceway to pond, until they enter the river or lake. But they will have got thoroughly used to the water, and a number will remain in the ponds and raceways for a while, taking up positions where there are eddies and suitable currents. It is a good plan to feed them artificially for a time in these nurseries, and some people approve of placing a fine movable screen across the bottom of the pond into which they are at first turned. This prevents them dropping down stream too soon, and will often cause them to head up again, should they find their downward course checked by it Under the care of a skilled fish culturist a large percentage of fish dropping down to the screen may be coaxed up stream again by judicious and careful feeding, and the screen may then be removed for half an hour, so as to allow any that still remain there to pass down into the nursery below. The fish often assemble in crowds about the embankments and below the pipes, where there is an under current as well as at the head and tail of each nursery pond. Many of the fish will not remain long in the nurseries at all, but never mind if they do not. During the short time they have been there they will have received sufficient education to be able to look after themselves. We know how they can take care of number one in a natural stream, where they have enemies at every turn.

I have often watched multitudes of trout fry dropping down our mountain streams in May or June, passing from pool to pool, from shallow to shallow, now hiding behind a stone or a tuft of water-moss, now passing on carefully but steadily, and feeling their way, as it were, all along the course. This they truly do, and in a much greater degree than may be supposed, for like the swallow leaving us in autumn and returning to the same locality in spring, many of these young trout return to the place of their birth. There is an instinctive knowledge implanted in a fish, which man in his civilized form does not possess, and the more we learn about them the more we find there is yet to learn.

Under proper conditions fry are excellent travellers, and, as a rule, two thousand of them will go into the same amount of water that would be occupied by about a hundred yearlings. This, of course, materially affects the cost. Their safety during transit in suitable carriers is almost an absolute certainty. There are, of course, contingencies which may arise *en route* which are sometimes distressing to the little fish, but these are found practically to be of very rare occurrence. Passing through a tunnel with the van windows open will fill it with foul air, some of which is taken up by the water. Two or three late passengers jumping into the van just as the train starts, and smoking there during the whole of a long run, also makes a very trying dispensation for the fish. These occurrences formerly took place, but now such large quantities of living fish are travelled over the railway systems of this country, that their requirements are better understood, and on most of the main lines I have usually met with the greatest courtesy, and had every reasonable assistance willingly rendered. The railway companies know their work, and as a rule do it, and their revenue is considerably increased by the live fish traffic, and will be much more so in the near future.

Fry travel better in glass carriers (carboys) than in any other apparatus yet discovered. I used them successfully many years ago, and travelled fish in them safely to and from America. Then metal carriers were strongly recommended by some fish culturists, and I tried them, but soon went back to the glass bottles, and my verdict to-day is that nothing has been found to beat them ; indeed, I have yet to see the carrier that will equal them for convenience, cleanliness, evenness of temperature, lightness, and durability. There is, indeed, nothing I know of that will favourably compare with them when properly constructed. I have them made to order, and they are so shaped that when filled to the proper level the angle of the water with the glass is such as to cause a continual and gentle splashing of the surface water only, during transit. The water in splashing is thrown against a canvas cover, and from this falls back again, and the slight vacuum which is caused is at once filled with fresh air drawn from outside, and so the work goes on, forming a self-aërating machine that is unfailing in its work as long as the train or cart is in motion. The splashing itself is confined entirely to the surface of the water, but the motion generated, owing to the shape of the interior, which has been carefully studied, sets up a rotary motion, and produces a perpetual current in which the fry enjoy life as in a brook, probably knowing little difference. I have travelled millions of fry in these vessels long distances, both by sea and land, and some-times under very trying circumstances, yet with perfect success.

Never under any circumstances should they be filled with water. It absolutely prevents any jar or splashing, and may prove fatal to the fish. Only once have I known such a thing to be done, and that was when I was once transferring yearlings a distance of about five miles, and used carboys for the work with perfect success, with the exception of one load, and in this case a booby of a carter, under the impression that he was giving the fish plenty of water, filled the bottles brim full, with the

I

very natural result that three-fourths of the fish were dead within an hour.

The weather in April and May is sometimes very warm by day, with frost at night, but in the carboys the temperature of the water is found to vary little. For short journeys small quantities of fry will travel very well in suitable tin cans, care being taken that they are not allowed to stand in the heat of the sun. When fry are turned out the operation cannot be performed too gently— indeed a great deal of the success of the work depends on this. The best plan is to use a clean bucket, into which the fry may be poured. Take care that the temperature of both waters is nearly alike. Should this not be the case, gradually raise or reduce that of the water in the carboy. Then pour some into a bucket, sink the latter gently in the water of the stream or tributary so as to allow it to fill with the least possible disturbance, turn it gently over on its side, and slowly withdraw it, practically "swimming the fry out." To do the thing perfectly I prefer floating a box, or several if necessary, and turning the fry into them, getting them on the feed and happy, and after some time, often twenty-four hours, withdrawing the screen and allowing them to escape. This process will be described more fully under the head of fry-rearing. It is dangerous to pour out fry suddenly into fresh water. The little creatures are very delicate, and are easily killed by a sudden shock to their systems.

I once took three thousand fry to a lake in three small cans, and on arrival there the fish were inspected by the members present of an Angling Association. They were in perfect condition, but whilst I was looking at one of the cans and replying to some questions, a leading member took one of the others, walked to the end of a boat landing stage, and uttering a short speech upset it, and its contents were discharged suddenly into the lake. "Why, they're all dead," I heard one of the company exclaim, and when I turned round I found fully two-thirds of the little fish on their backs, and they died. It was somewhat annoying, but the act was well meant. The members looked at me for an explanation, so I said, "Well, if you ask me to come here and turn out fish and then take the matter out of my hands like this, I am not responsible. Those fish have just been murdered." I then took the other two cans, each containing a similar number of fish under the same conditions, and gently turned them out without losing one of them. I mention this just to show how easily fish may be killed by a little want of knowledge on the part of the operator. There is now little excuse for such an occurrence, as everyone who takes an interest in the matter has abundant opportunity for becoming possessed of the necessary information to make the turning out of fry a perfectly successful operation.

Yearlings and larger fish require very different treatment. I have sent many of them in carboys, and they bear the transit well; but metal carriers are found to be more advantageous, as these fish are travelled at low temperature and ice is used, and a greater bulk of water being required, glass contrivances would not be so practicable. The carriers that are in use are, like the carboys, constructed on scientific principles, the convenience and comfort of the fish during a journey having been well studied. There are many different varieties of carriers, and I will describe a few of the most useful.

One that I have seen in use in the United States is a very simple affair, consisting of a tub having a wooden lid with a six-inch round hole in the centre of it. The lid being sunk a little below the top of the tub, any water which may splash out at once runs back again. The whole may be covered by a piece of perforated wood, zinc, or some netting, or by a funnel-shaped vessel in which ice may be placed. Another carrier that is also in use in the United States consists of a tin or galvanized iron can, bound round with wood, or fitted into a wooden case looking something like a cheese box.

In Germany a wooden apparatus resembling a flattened oak barrel is used, which rests on one of its sides, and the fish appear to travel in it very well. In the upper side is cut a square hole about six inches wide, and into this is fitted a wooden frame made with sloping sides, and a bottom covered with perforated zinc. Trout packed in these carriers are sent alive to market.

Oak casks charred inside make excellent vehicles for travelling fish, and I have used them most successfully; but the metal carriers commonly in use in this country, and made by

Messrs. Graham and Morton, of Stirling, are very convenient, and do not readily get out of order. They also stand the wear and tear of railway traffic as well as anything. Similar cans were first used in America, and I consider them about as good for travelling yearlings as properly made carboys are for travelling fry. I have also travelled two-year-olds in them very successfully, and on several occasions larger fish. When used for large fish, however, the work should only be done by those who thoroughly understand the matter, or the result may not be altogether satisfactory.

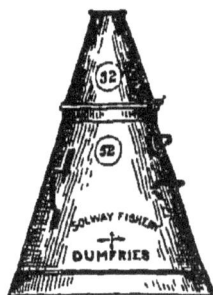

FIG 5.—FISH CARRIER.

I am often asked which is the best time of the year for turning out yearlings. In reply to the query there are many points to consider. I have turned them out every month in the year, from August to May, and carefully noted results; and the fish turned out during the latter end of August, or as early in September as practicable, have won the prize. They have made better fish the following summer than those turned out in spring. The only objection to the plan is that we often have warm weather just then, and, therefore, the cost of the work is much greater than it is a little later, say in November.

It is only in the care of experts that trout can be travelled in August, but when once safely introduced to the water, they make good progress. They are taken from the nursery ponds, where they have been herded together and have not room to grow as they might do, and are put into water where they have ample room, at a time when it is well stocked with natural food. They have been fed several times daily, and are, therefore, accustomed to having a good and regular supply of food. The starving they get whilst being transferred makes them feel hungry, and the consequence is that they begin to feed at once. They have come from crowded waters, where the natural food was all cleared out, and they were dependent on the artificial. They like the natural food better, and they eat freely and thrive upon it amazingly, getting thoroughly acclimatized before winter sets in. As a result they make good fish, and I have instances of their reaching half-a-pound in weight the following summer.

In 1887 I first noticed the great advantage that accrued from turning out trout yearlings in August and September, and followed up the experiment with a repetition of the same success in 1888. In 1889, on August 26th, I set out with 2,000 yearlings in a special railway car, fitted with suitable apparatus and a good supply of ice, and delivered them in safety at a station on the Highland Railway. The fish left the yearling house at the Solway Fishery at four p.m., and arrived at their destination at two thirty-five a.m., a journey of ten and a half hours, and on examination of the cans not a single dead or ailing fish was to be found—all were in perfect health. Here, however, a most unexpected delay took place. The fish were for a loch some four miles from the station, and the carts and men were waiting there, but the keeper in charge of them assured me that it was quite impossible to start until it was daylight, as there was no road a greater part of the way, and it would be quite impracticable to do it in the dark. I protested and declined further responsibility, but he remained firm, and would not order out his men and horses, and for three hours we waited. The morning was dull and drizzly, and daylight was slow in appearing. We had a comfortable room at the station, with a good fire and a liberal supply of both solids and liquids for the inner man, which had been sent down for our benefit.

I had many a look at the fish, feeling anxious about them, and at the end of three hours or so a few were showing signs of weakness. On being told that the fish were dying, and that I would have nothing more to do with them unless a start was made at once, the horses were yoked, the carriers stowed on the carts, and the journey commenced. The road was good enough for a mile and then lay over the heather, but some carts having been over a short time before with materials for a shooting hut, I found we could have managed it quite well had we started earlier. I had travelled a rougher way on a darker night, and saw no difficulty whatever. The road was rough and there were streams to ford that were rougher still, but the water was low at the time and we got safely through, though the jolting was very great, and the sickly fish at the surface were soon killed. Those that were in better condition headed down in the cans and were all right. Notwithstanding all the delay at the railway station we turned the fish into the loch with an actual loss of only some five per cent., which could have been prevented had we started at once.

At the lock-side, however, another delay seemed imminent. The keeper said the fish were not to be turned out until the purchaser had seen them, and on inquiry I found it would be at least two hours before he could arrive, and as events afterwards proved, it was over three hours instead of two. It was rather trying to stand there arguing the matter, knowing that every five minutes was of importance, and seeing the little fish already gasping in some of the cans. I soon made up my mind, however. I had gone there to turn out fish and not to stand by and see them murdered, so I got to work, and turned the fish out to save their lives, reserving a few, which we put into a small hole we found near, to serve as a sample of the bulk. The wait at the station I had no control over, but here with the loch before me, and the cans of fish standing at its margin, the case was different.

I have known several instances in which trout have been received in the evening, and left standing in the cans all night, with what result it would be needless to explain; and yet it is really necessary to say for the benefit of many that trout, if left standing in cans, will soon die. During transit the water in the carriers is in constant motion, and the fish are thus kept in a healthy and lively condition. The trying portion of a journey is the wait at the junction, and this has to be duly considered before the fish are started off, and the bulk of water in which they are travelled regulated accordingly, as well as the condition of the fish themselves. Trout in their normal condition will not travel; they require careful preparation for a journey, and according to the length of the journey so is the course of the preparation regulated. The pollution of the water by the fish themselves is one of the points to be carefully guarded against, otherwise it is most fatal. Some water in which trout live very well has been found absolutely unfit for them to travel in. The old plan of changing the water *en route* has been proved to be a very bad one, unless it be done under very exceptional circumstances and by those who understand it.

Towards the end of August and in September, 1890, I carefully turned out a number of yearlings, some of which were put into Loch Fern already referred to. I carefully watched the result of this experiment, which I conducted personally, and over which I took considerable pains. Early in the summer of 1892, some of the fish were taken running up to a pound in weight, and were the finest fish the loch had produced since its refilling. Other waters were stocked with fish from the same crop about the same time and later, and excellent results have in several cases been reported.

Further experiments were made in 1891, and from the results of these and a number of subsequent experiments I am satisfied that the earlier in the season yearlings can be introduced into suitable waters the better. It is much easier to transplant them in November and later, but at that time the water is colder and there is not nearly so much food present, and by waiting until spring a good part of the season is lost, and more than a season as regards the growth of the fish. Perhaps a better way of putting it is, that by turning out, say, early in September, a season is gained as against turning out in spring. Yearlings in September are from two to four inches in length, whereas in spring the Scotch yearlings of commerce run from two and a half to five inches. But the fish of September out-turning are found to have grown considerably beyond this size in a great many instances, and I have had them up to nine inches by the beginning of April. The reasons for this to anyone practically acquainted with the subject are apparent, and have been already explained.

There is yet another way of stocking waters, and though I refer to it last, it is by no means of the least importance. It is by sowing or planting "fully eyed" ova in artificial hatching beds, and in skilful hands is one of the best and most economical methods now in use. "Fully eyed" eggs are obtainable at such low rates from fish culturists that they can be sown in large quantities at a comparatively trifling cost. Care should be taken in the selection, and I prefer those which have been carefully incubated on glass grilles. They give better results. I have tried many experiments, and can produce much better fish, and hatch ova with a lower death-rate on glass than I can by any other method. I have, therefore, kept to the system, and can speak very highly of it. I believe very good work has been done by many of the other systems now in use, but where the best results are desired by all means use glass. I shall have more to say about this in another chapter.

There are at least three great advantages attaching to the use of ova, viz. :—

(1) They bear packing and transit well;

(2) They can be sent to any part of the world;

(3) They cost very little.

In searching for details relating to the success attending the stocking of waters, I find that the use of ova has played a very important part in the history of the world's fish-culture. The Chinese have been fish-culturists from time immemorial, and they deal extensively in ova, collecting and carrying to market the eggs of their fishes, and making them regular marketable commodities. The splendid results which have been achieved in New Zealand, and also in Australia and Tasmania, are due to the use of ova. Trout eggs were sent from this country twenty-five years or more ago, and the result is that to-day their waters are, in many cases, stocked with fish, and it is also a notable fact, that the trout have grown, in many places, to a greater size than is attained in this country. Ten to twenty pounds seems to be not an uncommon weight for *Salmo fario* in some New Zealand waters, whilst much greater weights are occasionally recorded.

The history of trout-culture at the Antipodes is very instructive. About 800 trout ova were successfully hatched in New Zealand in 1868, and these ova were obtained from the natural spawning grounds in Tasmania. Now, we find that the first introduction of trout into Tasmania was effected in the year 1864, being only four years previous to the introduction to New Zealand. During that year a small number of eggs were sent out from this country by Mr. Frank Buckland, Mr. Youl, and Mr. Francis Francis, the number being about 2,700 altogether. As a result of the importation of trout ova into Tasmania, and their cultivation, we find, in four years, that country sending ova, taken from fish

AN ANGLER'S PARADISE, SOLWAY FISHERY

on the natural spawning beds, to New Zealand. We find also that those eggs were successfully hatched there, and from this small stock a beginning was made, and there seems to be little doubt that from these eggs trout originated in New Zealand. So successfully was the work carried on there, that the New Zealand Government very wisely took it in hand, and the result was a considerable importation of ova into the colony.

Take the state of things in New Zealand to-day, and what do we find? Why, that the rivers of that country are, many of them, full of magnificent trout that have grown beyond all expectation. Trout-culture in New Zealand is a grand success. A friend, writing to me from Tasmania, August 7th, 1890, says: "The English brown trout that have been acclimatized here have done remarkably well, and attain a great size."

So then, in Tasmania also, trout-culture, though carried on under the great difficulty of importing ova from Britain at a time when its treatment was but very imperfectly understood, has proved a decided success.

In the United States the rivers of the Pacific Coast which contained no shad, were successfully stocked with those fish by transferring ova from the East Coast rivers. At first a million ova were carried in suitable apparatus, the incubation going on during transit. This proving a success, several cars were run conveying five millions each, and by means of these ova the rivers of the West Coast were stocked. The fish, which are prolific, multiplied very rapidly, and had become so plentiful that they were sold at three cents a pound.

In 1886, a quantity of the ova of the smelt (*Osmerus mordax*) were sent to Cold Spring Hatchery, on the north of Long Island. They were hatched and turned out in Cold Spring Harbour, and in two years a number of fish from these eggs were taken in Oyster Bay, which adjoins the harbour on which the hatchery stands, and into which they were turned, and they have also been seen in the streams.

Great success has in very many instances attended the planting of fry in the United States as well as in Canada. Had "fully eyed" ova been judiciously planted in artificial beds, probably the results would have been more satisfactory still. In

Canada, the magnificent river, Restigouche, flowing into the Bay of Chaleur, was depopulated, until the catch of salmon by anglers was only twenty fish in a season, and the whole commercial yield of the river was only 37,000 lb. weight. Hatching was commenced, and the yield in ten years was up to 500,000 lb.

The United States Fish Commission succeeded in introducing salmon (*Salmo salar*) into the Connecticut river, where previously it had disappeared for three-quarters of a century. In 1878 several hundred salmon, from 10 lb. to 15 lb. in weight, were caught running up this river, the result of fry planted there in 1874.

A considerable volume could easily be filled with accounts of the successful results attending the stocking of waters, and in our own country we have many cases in which the most satisfactory results have accrued. As I write, by the side of a natural trout stream, I can see the trout disporting themselves in numbers, nearly every fish in the pool before me being the result of artificial culture, whilst in an artificial stream close to, on which are many deep and spacious pools, and where the fish are fed, large quantities of magnificent fellows up to several pounds in weight may at any time be seen. Over the hill in the next valley is an artificial lake, which is well stocked with fine trout, nearly all of which have been artificially bred, and beyond this other lakes, reservoirs, and ponds, all well stocked with magnificent fish. If we go further afield we have Loch Leven, the statistics of which, extending over many years, are strikingly in favour of fish culture.

In Wales, too, we have the well-known Lake Vyrnwy, the fish supply in which is now kept up by a well-ordered system of artificial cultivation. The successful introduction of grayling into the Nith and many other rivers is another proof, if any more be needed, to say nothing of many Highland lakes which are now well stocked with trout, where, in some cases, no trout were before.

Of the success of trout culture there can be no dispute, and I maintain that what can be done with trout may be done on a far greater and more profitable scale with salmon. That is a point about which I am quite convinced, and I would carry it further and apply it to many other fishes than those belonging to the

Salmonidæ. There is, however, this difference, that trout being retainable in fresh water ponds can be successfully cultivated by the individual, whereas salmon must be allowed to go to sea if they are to produce the highest results, and this renders individual action somewhat impracticable. By a well-directed system of co-operation amongst owners of fisheries, it is beyond any doubt that splendid results may be obtained. Of one point I have no doubt—that no investment would pay a much better dividend if properly managed.

It has often been stated by scientists and others that only about one trout or salmon egg in a thousand deposited in our streams, produces a mature fish. This, probably is not far from the mark. Anyhow, we are quite sure of one thing, and that is, that the rule applies to over ninety-nine per cent. of the ova deposited naturally in our streams; seventy-five per cent of this loss probably occurs before the eggs are hatched, and during the hatching period. It will be apparent at a glance, that by taking charge of the ova and actually hatching over ninety per cent. of it we are doing good work. It is necessary, however, that it be done properly, and that is just what has often not been done in the past, and cases of failure which the practised fish culturist could foresee, and which were inevitable owing to the means employed, have tended to bring fish culture into bad repute. Circumstances are entirely altered now, however, and the facilities which are provided for sowing good well-eyed healthy ova in our waters, will ere long produce good results if properly utilised. There are some individuals who still assert that fish culture is a failure. So there were those in years gone by who pronounced the steam engine a failure. Anything that is not absolutely perfect in all its details is pronounced a failure by a certain class of individuals, and probably always will be. Fish culture, however, as applied to the *Salmonidæ*, has been proved by the results to be a great success, and I venture to say that in the future it will be still more so. It is now being successfully applied to the growth of other fish, both marine and fresh-water, and as information is gained by experience, and difficulties are bridged over, its practical use will be found to be of great service in the management of our fisheries.

CHAPTER V.

THE HATCHERY.

FOR the benefit of those who wish to do their own work from the commencement I will endeavour, in as few words as possible, to describe the various needs and processes of the practical fish culturist. The first thing essential is a hatchery of some kind. It may be large or small according to the amount of work required to be done, and may be fitted up in different ways, but although the details may vary, yet the principle of construction is the same, whether small or large. It must be near a good supply of pure water. I do not mean chemically pure, but naturally so—that is, it must not contain any excess of mineral matter of any kind, and it must be free from mud or sediment. The water which flows from a good clear spring and is wholesome to drink is usually good. But the best way of proving it is by means of the fish themselves. Do trout frequent it, and do they spawn in it freely? If they do, it is probably all right; if they do not, then be careful, and should it seem clear, on examination, that they avoid it, then be very cautious in using such water. By all means have it analysed, and find out exactly what it contains, and what it does not, before commencing work. Brook water is the best for growing the fish, but spring water is usually acknowledged to be safer for hatching the ova, and chiefly for two reasons—regularity of temperature, and freedom from organic and mineral matter in the form of sediment.

VIEW IN MAIN HATCHERY SOLWAY FISHERY.

Suitable water is such an important factor in the successful working of a hatchery, that too much caution can hardly be used in the selection of the site for the building. Some very clear and good-looking waters are not good, and it really becomes the work of an expert to decide what is suitable and what is not. I have seen excellent work done in a hatchery where only river water has been used, and I have seen spring water that to look at appeared perfection itself, yet did not do its work at all satisfactorily. Some spring waters contain too much iron, lime, or other deleterious ingredient, and hence the great care that is required in the selection of a suitable supply.

Therefore, where the incubation and hatching of ova is to be carried on on an extensive scale, it is better to consult an expert.

Where limited operations only are intended, test the water by keeping some trout in it, and, if possible, hatch a few ova, and rear the fry for a season by way of experiment.

Have the water analysed.

Having selected a suitable spring, the next consideration is the construction of a hatchery. I have seen several sets of hatching apparatus worked out in the open air. The objections are that such hatcheries are exposed to the action of frost, which in very severe weather is likely to cause damage, and they are liable to be tampered with by man or beast, which should not be the case. I once had a spring which threw a copious supply of excellent water, never below 38° Fahrenheit. Such water would answer well for an outdoor apparatus, as it could easily be made to pass through a series of hatching boxes before being reduced to the freezing point. But most water would be likely to give trouble at times. Therefore, if it be practicable, place the hatching apparatus inside a frostproof building. In our climate an ordinary stone and lime wall is sufficiently frostproof for the purpose; an ordinary slated-roof is not. Thatch will do, but it has the objection that it needs constant repair and harbours vermin. Underground hatcheries, when the situation will permit their construction, are excellent. But an ordinary stone-built and slated building will be found in practice to answer all requirements if felt be laid under the slates. The temperature inside it may be kept at any desired point by means of hot water pipes, which answer well, and prevent any mischief during the severe frost. The heating apparatus should be outside. Never have a stove of any kind in a hatchery. I was once persuaded by a man who "knew everything" about fish culture to try one, and never was a greater nuisance. However well kept, smoke would at times escape, and anything of this sort is to be carefully avoided. For the same reason I have found it necessary strictly to forbid tobacco smoking in the hatchery. I did not do this at first, the building being large and well ventilated, but carefully watched the effect of it on the alevins, and found it very hurtful.

Lighting a hatchery when work has to be done, as done it must be during the dark hours of winter, is a matter that requires the greatest care. Oil lamps of any description are to be most carefully avoided. I have never from the first allowed anything to be used except candles (not tallow), with the exception of the watchman's bull's-eye or other lantern when on his rounds during the small hours. Even this, although most carefully used, and according to strict rules, was found to give trouble. A very small drop of oil may do harm should it get into the water, and where oil is used there is always a danger. Candles only are now allowed in the hatchery, and are found to work well. They are carried on simple wooden candlesticks, each made to hold three candles. These give enough light for the laying down of the ova, which is almost invariably done after six p.m. Occasionally candle droppings may get into the water, but as they float and immediately solidify, they are quite easily picked out again, and I have never found them do any harm.

Except when there is a great press of work, spawning is not, as a rule, commenced before ten a.m. Fish spawn better later in the day when the temperature rises a little, and the eggs taken in the afternoon are carefully washed and placed in bowls in the hatchery, ready to be laid on the grilles as soon as the spawning operations are over for the day, and the fish removed from the spawning tanks.

The water should be brought into the hatchery from the spring in glazed earthenware socket and faucet pipes. The joints should be well cemented, and the pipes laid underground. It will probably require filtration, although sometimes it is sufficiently pure to do without. But this is a very rare exception. At or near the point where it enters the hatchery then, construct the filters. For pure spring water half-a-dozen or more flannel screens will usually be found ample, and often three or four will do. It is better to have too many than too few, for they play a very important part in the success of the work. For fifteen years I have worked with no other filter, and now that a much larger volume of water is required in the hatchery the same method is essentially successful, except that the water is first passed through a couple of settling tanks, which are found very useful adjuncts.

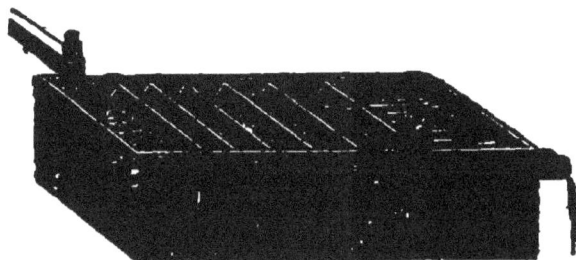

Fig. 6.

A simple filter shown in Fig. 6 explains itself. It consists of a wooden box, six wooden frames with coarse flannel stretched on them, sliding into groves at a moderate angle, an inlet and an outlet, and the whole charred inside. The size depends entirely upon the amount of work to be done and the state of the water. As an example, I may say that I have incubated successfully half a million ova in the water discharged through a set of four (occasionally increased to five) flannel screens, of about seventy square inches each. A double set of these (for convenience in cleaning), each in a separate box, is used, the whole water sometimes passing through one box, but, as a rule, both boxes working.

The filter boxes I have at present in use are twenty-four inches by twenty-four inches, and two of these boxes now working will pass 200,000 gallons of water per day, or enough to incubate four millions of ova. Behind each filter are two settling tanks built of concrete. This applies to the main hatchery only, two other buildings which are used as accessories having an additional supply.

By way of caution to beginners I would say that concrete should in all cases be very well seasoned by use, before allowing the water passing over it to be used in the hatchery. All work is done at the Solway Fishery, as far as possible, a season in advance, or in early summer, and by allowing the water to run for a few months, everything is rendered perfectly safe. I have seen places where the work has only been finished the day before the eggs have been laid down, and where the water supply has actually had to be cut off afterwards in order to rectify little matters that had been overlooked. These are the sort of places that bring discredit upon fish culture. A fish hatchery and everything about it should be clean and sweet as a dairy, and should be kept so, and on this largely depends its success or failure. Good ventilation is essential, just as it is in a house for growing plants. Too much light should be avoided, and especially large windows facing the south, which would let in the glare of the noon-day sun. It does not matter much how the place is lighted, if attention be paid to these points. It may be by sky-lights or by side windows. I commenced work thirty years ago in a conservatory, which is now about the last place I would choose for the purpose, but yet I got on very well.

The floor of the hatchery should be of concrete, or of stone, or suitable pavement of some kind. Whatever material is used, take care that it is rat-proof. The level is also a matter for con-sideration. Some fish culturists advocate dry floors, the waste water being carried off in pipes, and drained away underneath. This may be all very nice, but in a working hatchery it does not answer very well. There is no harm in having the floor wet, if the house be properly ventilated, and as water must often be spilled or even emptied upon it, it is, for several reasons, better to have open gutters under the hatching boxes than drains laid underneath the floor. There are two great objections to such drains. One is, that if anything should happen, and they have to be examined, it necessitates the pulling up of the floor, and should anything go wrong in the middle of a hatching season it might be a very awkward matter; another is that drains are apt to encourage rats and foul smells. There are also many other inconveniences. I have tried both systems, and I am greatly in favour of open gutters on the surface. They are simple, and add but little to the cost of the hatchery; they are convenient and clean, and are always open to view.

The floor itself, I need hardly say, should not be level, but should have a fall one way or the other to suit circumstances, and to cause all water to run off immediately into the gutters. It should be frequently washed, not with a floorcloth, but with clean water and a broom. This should be done by the manager himself, or by someone working directly under him, as it requires to be done with care. On no account should any carelessness or undue roughness be tolerated in a hatchery. The disarrangement of a pipe or a tap, or a blow on one of the hatching boxes, may do serious damage, and whoever cleans the floor should be acquainted with the working of all such contrivances.

The door of a hatchery should be kept shut, or rats, mice or birds will get in, and it is sometimes difficult to get them out again. Rats seem naturally attracted to a hatchery, and if they can get in they will. The outlet for the water should be carefully guarded by a grating, and every precaution taken to prevent any intruders of this kind gaining access to the hatching boxes.

Where lead or iron pipes are used for bringing in the water, it should be borne well in mind that either may be very injurious. I have seen excellent work done with them, and I have also seen great destruction caused by them. The first hatchery I ever built, which was erected in one of the suburbs of a large town, was supplied from the waterworks at the rate of sixpence per thousand gallons. The water was excellent, but occasionally repairs were done by the company, and on these occasions it would for a short time run very thick and yellow, and would poison my little fishes. In the same way when the water has been shut off a hatchery for a few months during the summer the pipes would corrode, and afterwards send down a quantity of poisonous matter, which might do great injury. Care should, therefore, be taken that they are thoroughly clean before hatching begins. This applies not only to the supply pipes, but also to the hatchery itself, and all the apparatus which it contains.

K

So much for the house—now about the furniture. A glance at the accompanying illustration will give a general idea as to what is required. First of all, two distributing tanks. These are two long wooden boxes or troughs which receive the water from the filters, one being used for distributing the spring water to the hatching boxes, and the other being used for the brook water. Place them overhead if practicable, but in cases where the water cannot be got up to that level, they may be placed three or three and a half feet above the floor. In the latter case a series of round holes one inch in diameter, and short tin or lead pipes four inches long will be all that is needful for supplying the hatching boxes. Bore a hole, and fix a pipe so as to deliver the water into the upper end of each hatching box or set of boxes, which must be placed just below the bottom level of the distributing tank. Regulate the supply by a small piece of tin or wood, sliding in a groove made by nailing two slabs on to the inside of the tank. The tanks themselves may be nine inches wide by nine inches in depth.

The advantage of placing the tank overhead where practicable is that it allows the operators to pass along that end of the hatchery, which is a consideration when each range of boxes is close on eighty feet long. The outlets for supplying the hatching boxes can then be made in the bottom of the distributing tank, and by using short pieces of lead pipe, each with a flange a quarter of an inch from the end, for nailing to the inside bottom of the distributing tank, a screw-tap can be attached, which is a great convenience in regulating the water, and on the whole better than the small-scale sluice already alluded to. Fix a short piece of indiarubber hose pipe on to the tap and the whole is complete. Take care that all is well seasoned before using the water.

The hatching boxes are very simple contrivances, but require to be properly made or they may be found not to answer their purpose satisfactorily. A very good size to make them is twelve feet long, by nine inches wide, and six deep. Near the inlet end fix a board across the box at a slight angle, and reaching down to within an inch of the bottom. This serves to break the force of the water and prevents it from washing the eggs off the grilles. At the other end, in the centre of the bottom, and two inches from the box end (see A FIG.7), bore a hole one and a quarter inches in diameter, into which fit a plug. The end of the box is six inches deep like the sides. Take a saw and cut into it two inches from each side. This leaves five inches between the saw cuts, which should each be three inches deep. Take out the piece of wood between them. Make a wooden outlet spout four to six inches long, and fit to the opening, taking care to give it a slight fall outwards, which will cause it to throw the water well into the box below. When the outlet spouts are fixed level, and the water supply slackens a little from any cause, the bulk of it will often be licked back underneath the spout, and so will fail to enter the other box (see Fig. 8). This endangers the eggs in the lower hatching boxes, and there should be no possibility of such an occurrence taking place, as the consequences may be serious.

Fig. 7.   Fig. 8.

I have recently doubled the width of all my hatching boxes, chiefly for the sake of economising space, as the double boxes really take up less room and do more work. The thickness of two sides and the space between them is saved, and more eggs can be hatched in a box, it being more roomy and having two currents of water. This is a decided advantage, for owing to one current acting with the other, eddies and counter currents are produced which did not before exist. The necessity for extra rearing boxes is done away with, as the hatching boxes are quite sufficient for all purposes, if the fry be turned out any time within fourteen days of commencing to feed. In a large hatchery where millions of ova are incubated space is an object. In a small hatchery the arrangement does not so much matter, and must depend on local circumstances, such as the shape and size of the building, etc.

Each hatching box must be provided with a screen, which consists of a charred wooden frame covered with perforated zinc (No. 7 zinc is a good size). This screen slides in a groove, made by cutting a strip out of the sides of the hatching box as shown by the dotted lines B c in Fig. 7. This groove is made, of course, on the insides of the box, and should be carefully charred. A piece of flannel should be placed between the screen and the box, and great care must be taken that it fits perfectly tight, and that no newly-hatched fish can get through anywhere. If it be by any means possible, most assuredly they will. Rests require to be put in along the insides of the box for the grilles. I have tried many plans, but find nothing better than small galvanized staples, which take up little room, do not make corners into which the fish can

Fig. 9

Fig 10.

Fig 11.

get, and are not much in the way at any time, when not occupied by the grilles. They should be varnished. I have used inch wire nails driven in half of their length, but I like the staples better.

All wooden hatching apparatus should be carbonized wherever the water comes in contact with it. Elsewhere it may be painted. The carbonizing, or charring as it is commonly called, is done by working hot irons over the surface of the wood. The bigger the iron the longer it keeps hot, but the greater amount of heating it requires to make it hot enough. It often happens that at a hatchery a very large fire is inconvenient, and, therefore, it is better to have comparatively small irons. To begin with, procure from an ironmonger two of the largest ordinary "flat irons" that are in general use. They will be found to do the work well, and are very handy. Should it be found desirable to have something larger get one made like Fig. 9. A hook can be placed in the loop with which it is furnished, to lift it from the fire and drag it along by when being used.

For charring the grooves Fig. 10 will be found very useful, and for the plug holes I use an iron shown by Fig. 11. The object of charring the wood is to prevent the growth of a fungus (*Saprolegnia*) that is very deadly amongst ova and fish. It grows vigorously on wood, but will not grow on carbon. By thoroughly carbonizing the apparatus, therefore, a great danger is to a considerable extent averted. The process should be carried out by a steady-handed careful man, as the wood requires to be very evenly burnt. The carbon wears off in time, and each season I give my boxes a coat of black varnish, and sometimes two. Some fish culturists repeat the carbonizing process each season, but the varnish is by far the most economical, and answers quite as well after the former has been once done.

In the charring the heat opens the seams of the boxes, and at times causes the wood to crack, and it certainly has a tendency to make the joints leaky. The varnish has just the contrary effect : it fills up small crevices, and tends to make the boxes watertight, which is a great advantage. I got mine first from the United States, where it is largely used, but have since found out a way of making a varnish which answers admirably. There are many varnishes in the market which will do, but as the use of some of them is at times attended with danger, I shall be glad to supply anyone with that used at the Solway Fishery, which I have found to be perfectly safe.

At the head of each hatching box a board is fixed in a slanting position to act as a breakwater. About an inch of space is left underneath it, and the water is thus directed under the grilles, which should be about one and a half inches above the bottom of the box. Slates do very well instead of boards, and it will be found very desirable to have a pile of them at hand for this and similar purposes. They are easily worked to any size or shape, and are very useful about a hatchery. They are clean, and give off nothing, and need neither charring nor varnishing. A number of small water boards will be required for placing in the outlet spouts when it is desired to increase the depth of the water, and pieces of either wood or slate will be found very useful for the purpose. Both inside the hatchery and outside in the raceways, they are often very useful indeed. A trap box should be placed below the spout of the last hatching box in each set. It does not matter how small it is, but one side should be of perforated zinc, to within say an inch of the bottom. Through this box the water should pass before it is allowed to flow into the gutter that carries it off. The object of these trap boxes is to catch any fish that may be escaping, and they play a much more important part in the working of a hatchery than many persons would suppose.

I have more than once been told by one of my friends that his fry were decreasing in numbers, and he could not tell what was taking them. On my suggesting that they might be escaping, the idea would not be entertained for a moment. Impossible! When my friend, acting on my suggestion, had placed a trap box below his outlet spout, he soon found out where the fish were going. A few hours would reveal the fact that a dozen or more had escaped from the box above during that short time. On a search being made for the crevice through which they had escaped, it would be very difficult to find it, and when found my friend would hardly be convinced that the little fish could possibly have escaped through such a small aperture. Even in a well-ordered hatchery there is always a liability that some such occurrence may take place, and it is better, therefore, to be prepared for it. With these traps in use such a leakage would be detected at once.

There is another very important accessory to the hatchery also, and that is a catch pool. It consists of a pool or pond outside the hatchery, in any convenient situation. It may be near or at a little distance, but it is at least better to have it a few yards away, so as to get the advantage of having a raceway leading to it. It may be made of any convenient shape, but for practical purposes a long and narrow pond suits best. Some fish will be picked out of the hatchery boxes occasionally, apparently nearly dead. They are at least in such a plight that any fish culturist would condemn them. Do not throw such away, but as long as life lasts give them a chance. This is easily done, by putting them into the gutter which takes the water away from the hatching boxes. The writer was surprised, the first season this was tried, to find how many of them revived and made good fish. It will be easily understood that amongst a number of little fish in a hatching or rearing box, a few are liable, like chickens or any other beings, to become sickly. If left where they are these soon die, but by being given freedom, and plenty of room and water, and turned into a pond which is densely planted with suitable aquatic vegetation, many of them recover, and the wholesome natural food upon which they live soon makes good fish of them.

All hatching boxes should be provided with covers or lids, These should be made as light as possible in weight, and half inch boarding does very well. The eggs are much better for being in the dark; in fact light is bad for them. Be careful, then, to have lids for all the hatching boxes.

No one should be allowed to enter a hatchery but those whose business it is to work in it and visitors who desire to see the hatching and other operations. The latter should always be conducted through the building by the manager, or some other responsible person, who will be able to describe the various processes, and to see that nothing is disarranged. There is a latent propensity in human nature that from time to time shows itself in certain individuals, leading them, if they see a tap, just to turn it, or to try if a sluice be fixed or movable. These little acts, simple as they may seem, may cause incalculable damage in a hatchery. The same applies amongst the ponds. I once had a number of connecting rods projecting above the water, for the purpose of working some of the pond plugs, and quite a number of visitors who carried walking-sticks or umbrellas, gave these a tap in passing, and caused disarrangement of the outlet valves. I got over this difficulty, as I thought, by shortening the rods, so that the projecting ends remained some few inches below the surface of the water. Soon after this had been done, I was showing some visitors round, and one of them, seeing the end of a rod underneath the water, gave it a vigorous poke with his umbrella, asking at the same time what it was for. I showed him what he had done in disarranging the outlet valve, and he was profuse in his apologies. But how much better it would have been to have let things alone at first.

I once knew a visitor to poke a hole through a perforated zinc screen, with the result that a quantity of fish escaped from a stew, in which the owner had placed them at considerable expense. One day two individuals came to look at my fish, and without asking leave went amongst the ponds, and before they had been many minutes they were actually leaping over the fry nurseries, to save going round a few yards, and running a great risk of doing irreparable damage. This was rather too much, and they were, as a somewhat natural consequence, ordered off the ground. This they did not like, but who was to blame? Every part of a fish farm should be kept strictly private, and no one allowed to wander about without a conductor.

Before leaving the subject of the hatchery, I must again refer to the water supply. It is the driving power of the works, and must on no account be allowed to fail. Take care, then, that the supply is ample. There should always be plenty to spare, and this should be duly considered before entering upon the work. Another thing which is of the utmost importance is that there should be no possibility of the supply being by any means cut off. All the arrangements are now very perfect at the Solway Fishery, but it was not always so, and consequently I have had considerable opportunity of gaining experience. When I first commenced work, with no other building than a hatchery in a very isolated position, I soon found how absolutely necessary it was to be on the spot myself. I therefore fitted up a room in one corner of the building, with a cooking stove and a berth, and, having been accustomed to camp life, thoroughly enjoyed being lulled to sleep by the music of the water as it passed from tank to tank, and in the middle of the night awakened at once if a change took place in the sound, owing to the alteration of the currents. A whole winter was spent in this hatchery, the day being occupied in poring over the hatching boxes, watching carefully the development of the embryos and the growth of the little fish, whilst working amongst them, and the long winter evenings were spent in writing up my notes, reading, correspondence, &c. I cannot but look back upon this period of my life as one of pleasure and delight, varied as it was by an occasional day's shooting, or a hurried run to some part of the country, to inspect a lake or give instructions for the construction of a fish pond.

One of my friends, who occasionally came over from a smoky Yorkshire town, was wont to describe the life as a "continual pic-nic." Anyhow, the days went pleasantly by; there was plenty to do, and doing it was simply an enjoyment, and when the sun shone and all went well, I could occasionally leave the place for a week's cruise in my little yacht *Wildwing*, exploring the mysteries of the deep, and collecting marine specimens, &c. Do not let the tyro run away with the idea ·that there were no drawbacks, however. As life in general is said to have its "ups and downs," so had this life, which, though appearing to outside observers all sunshine, often carried with it a great deal of roughness and hardship. These were times when my friends of the sunshine were absent.

I have spent, during a severe frost, night after night amongst the ice, preventing the water supply being cut off. Every few minutes it would freeze up if not attended to, the floor meanwhile consisting of one frozen mass, while icicles hung from boxes, filters, and distributing tanks, reflecting back the light and making quite a pretty sight. Often when the sun got up and the frost slackened, and I could turn into my berth to get a little rest, I would take a last look over some of the hatching boxes to see that all was right, and at the sight of the crowd of moving little beings within, exclaim, "Yes! it's worth it all"; and after my sleep would rise refreshed, and just as ready for another battle with the Ice King as ever. One great advantage I had in these battles—and they really were such—was that I always came off victorious.

All this is altered now. The water, instead of coming into the hatchery through a long run of wooden spouting, several feet above ground, is conveyed for a good way in pipes beneath the surface, and other precautions are taken whereby freezing is prevented, and prevention is much better than cure.

Possibly some of my fish-cultural friends will laugh at the idea of allowing the water to freeze at all, but I know that more than one of them has been troubled in the same way, and have heard of water being found in a morning absolutely stopped by the frost. Such a thing was never allowed to happen here. I do not mean to say that a tap has never been accidentally stopped— far from it ; but such an occurrence was extremely rare, and the cause mostly unusual. Accidents have, however, happened, and for the benefit of others I will mention some of them.

All the hatching boxes are supplied with water by means of taps, as well as the yearling boxes in the yearling house. They are most easily regulated, and are safer as sources of water supply than any other apparatus I have had brought under my notice, but they are not infallible. On one occasion I remember a tap suddenly ceased to work, and nothing would induce the water to resume its course through it. Now all the taps are so arranged that in less than half a minute they can be detached by turning a screw, and in as short a time replaced, so that no time had to be lost in examining the one in question. In it was found a frog ¦ How it got there no one could ever tell. Like the one said to have been found in the middle of a solid rock, the manner in which it came there was a mystery which none could solve. The length of time it had been embedded was a much simpler question, for it must have been discovered the moment it succeeded in getting into position. A leaf has twice accidentally got into a tap and stopped it, but only for a moment, and on one occasion, while the screen was out for a few minutes, a truant trout of four and a half inches came down the spout, but unfortunately did not succeed in stopping the tap, for it passed right through it, and was only discovered, after lapse of many days, by the serious diminution in the number in a box of Alpine char. The trout (a wild fish) had the usual disposition to hide himself, and the box being against the wall of the hatchery, could only be examined from one side, and owing to this, he had remained unobserved until the mischief was done. When found he was simply gorged with char.

These occurrences only show what great care is required in a hatchery, and point to the necessity for having everything properly arranged in thorough working order, and being continually watchful to see that they are kept so. Without this care success is impossible. Having water entirely under control means security to the works, and is one of the first and most important considerations in connection with a hatchery.

For work on a small scale one of the most useful arrangements is shown by Fig. 12. There are many modifications, but the principle is the same. It may be worked against a wall as shown, or out in the open, the boxes being supported by a wooden frame. It may be inside a building or otherwise. One of the most important points to be observed in its construction is that the outlets are efficient. Be sure that they are capable of conveying all the water that will ever be required, and take care also that the screens are large enough. I have so often seen screens made by simply nailing a piece of perforated zinc on to the inlet end of a spout, that I feel it necessary to give this caution. A very simple way of making a screen is to nail three strips of wood, each of about six inches long, so as to form three sides of a square enclosing the outlet, and nail the perforated zinc on to them. I use these small hatching boxes occasionally for experimenting, and in my own I prefer to have the screen the full width of the box, fitted as described for the larger boxes. The results are much more satisfactory, as full-sized screens do not so easily get clogged, and when the egg shells are about this is important. There are other reasons also, which will be referred to in another chapter.

Fig. 12.

CHAPTER VI.

COLLECTING THE EGGS.

The old method as employed at Troutdale Hatchery—Ova hunting in Cumberland—Work on a natural stream—The water ousel—Blank days—Honister Crag—Ullswater—Advantages of the present system—Spawning trout—Laying down the eggs—Embryology—Dry method of impregnation—Catching the spawners—Sorting—Cleanliness—Effects of temperature—Washing the eggs—Hermaphrodite fish.

IT is now rather more than a century since the re-discovery in Europe of the art of fecundating and hatching fish ova by artificial means, but for many years the matter was only understood very imperfectly by a few persons, and was looked upon as nothing more than an interesting scientific experiment. For a long time it was supposed that the gravelly bed of a stream was necessary for the successful hatching of the ova of *Salmonidæ*, and the earliest form of hatching apparatus consisted of nothing but boxes with perforated sides, through which the water could flow, and which were sunk in the stream, filled with gravel and ova, and in due course, in many cases, some of the ova hatched. Even when I commenced the propagation of trout in this country thirty years ago, little comparatively was known of the proper methods of dealing with the ova and the young fish which they produced. But now all this is changed, and the ova can be taken and properly impregnated, and more than that, they can be properly incubated and hatched, and the delicate little beings known as "alevins" can be grown into large fish, as surely as the gardener can from his seeds produce a rich crop of flowers or vegetables, as the case may be.

I well remember in my early days the keen enjoyment experienced in hunting the mountain streams for the various breeds of fish, destined to become the ultimate producers of a far superior race. Many a time I have started off at four or five o'clock on a November morning for a long and tedious walk over the Cumberland mountains, often rendered even dangerous by the accumulations of snow and ice met with at that season of the year. After fishing all day we would come back tired and weary at night, with perhaps a few thousands of trout eggs in the collecting cans, and often enough with none. These were laid down carefully on the grilles in the Troutdale Hatchery, where the work was carried on for fifteen years. There was an amount of enjoyment in it which it is impossible to describe—it must be felt to be understood—and an excellent opportunity was afforded of studying the habits of the various species or varieties of fish with which we came in contact.

The information gained from practical sources in those days has proved of very great value, and has been of material assistance in the building up of a successful fish farm. It is now so extensive that it has been found quite impossible to get a good photograph of the ponds, but the accompanying illustration will give some idea as to the way in which they are laid out. I often smile as I remember the time when we hunted the trout in the wild mountain glens of Borrowdale and the neighbouring valleys in Cumberland, when, although armed with permission from riparian owners, we were loudly denounced by a certain class, many of whom ought to have known better, as poachers, &c. But out of it all has been acquired a mass of information which has enabled us to carry the work forward until it has assumed its present proportions. Instead of collecting ova from the natural streams, which at best is very arduous and costly work even when properly carried out, they are now taken in enormous quantities from fish reared in well-made ponds, which are entirely under control. Upwards of a quarter of a million trout ova have been taken in one morning, and of coarse fish I have taken a million before breakfast.

I have made these preliminary observations, in order that the uninitiated may at once realize to some extent the altered position in which fish-culture stands to-day, as compared with its position a quarter of a century ago. The eggs now obtained from the domesticated fish referred to require to be built up for months beforehand in the ovaries of the fish, and great attention has to be bestowed upon their proper development during this period, the fish, like other valuable farming stock, requiring to be very well looked after, the sexes kept separate, and the food varied according to season and circumstances. It is a matter that requires much skill and judgment, and no novice can expect to rear breeding fish successfully at first. I say breeding fish, for it is quite a different and an easy matter to rear fish for the table or for angling purposes. This can be done, and is now being done extensively, and I am glad to say, successfully. It seems strange that we have been so long in realizing that the discovery made by two obscure French peasants, that the ova of the trout could be fertilized and cared for, might be turned to good practical account, largely multiplying the number of fish in our waters.

In the early days of my experience as a trout culturist, when we were dependent upon natural streams for supplies of ova, a careful watch had to be kept upon the trout as spawning time approached. About the end of October, when the autumnal tinted leaves begin to fall in showers as the winds sighs through the grove, and the mountain tops have already been capped with snow, the trout begin to leave the larger rivers and streams, and push their way into almost every little rivulet on our hills, and in our valleys, in search of suitable places for making their nests and depositing their ova, and places unfrequented by them except in their earlier stages at other times of the year are then often found to be full of fish. Notwithstanding all our pains the trout often succeeded in eluding us, as floods would come down and prevent any work being done in the streams for two or three days, during which time the fish had spawned and perhaps even left the locality, for they do not remain long in the small tributaries; while there they are in danger, and they know it.

Let us take an imaginary ramble to a wild mountain glen. The little stream which tumbles over the rocks and boulders empties into a lake, which is full of an excellent breed of trout. As we near its mouth three herons rise and flap lazily over the placid surface of the lake, and a flock of wild ducks noisily departs from the reed bed in which it had been hiding. A little further on another heron leaves the brook—they know full well the trout are on the move. Passing on, we notice the solitary footprint of an otter on a little patch of yielding sand amid the stones; he was not discreet enough in his work, and has left a tell-tale behind him that reveals the secret of his presence. We round the corner of a mighty rock and push up-stream, soon coming to a waterfall. Surely no fish can pass this obstacle! Yes, they do, though; for when the flood comes down and threatens to sweep everything before it, the fish go up and reach their nesting ground far up the valley. They may have already reached it, and we may be too late. Passing along the rocky bed at length we reach a pool, and peering cautiously we see—yes! there they go—one—two—three fine trout, right off a gravel bed, down into the deep recess amongst the rocks. Put in the net; now close it round and set it properly, then with a willow sapling gently move the fish. There! two are in the net; now lift; we have them—beauties—each a pound, and females too. Well done! we'll try again. But, no; the other has "holed" and will not move; so leaving him we wander on. Soon we sight another, and at the first alarm under the bank he goes. We set the net and drive him out, but in he goes again, under a shelving rock this time. Again we try to poke him out, but it is of no use; and after several fruitless efforts we go on and try three other pools, but do not get any fish from them.

But look! what is that? A fish? No; a bird, emerging from the water, which scatters from its feathers as it flies, and swiftly disappears round a bend. We cautiously creep on and see him sitting on a stone, a little beauty, with a throat white as the driven snow. His tail perked jauntily, he faces us—he turns—he dips his body. Is he curtseying? No, not exactly; it's only his way. But there, he's in again, taking the plunge right merrily. How long he stays beneath! Yes, he's at home beneath the water and keeps us waiting in suspense. At last he re-appears with something in his mouth. What is it? shout! he's off and dropped it, and we run to pick it up. It is, we find, not a trout egg, but a water insect—a deadly enemy to the ova of *Salmonidæ;* and that is the work of *Cinclus* in October and November. He is one of Nature's workers, and he does his duty, saving many lives by taking others; and if he takes a fish in spring to feed his callow brood, perhaps he makes amends by keeping down their enemies at other times.

But to our work. We try another pool and get a fish, and

presently another. Strange both are females, and yet not strange either, for the males are higher up, and we often find it so. Further on we go and get another fish—a female too, and ripe, and working hard for more we take at length a couple, one of them a male. But the clouds are gathering on the mountain tops, and the pass is getting choked with mist and threatening snow. A storm is brewing, so we must be quick. There! we have another male, and letting this suffice we hasten down to lower grounds. The storm comes on apace. The raven croaks above, as wheeling in mid-air he bids defiance to the storm. We hurry on, but before we reach the valley it has burst upon us. There! we cannot do better now, so we will stop and spawn our fish under an overhanging rock; then, having milted the ova, get our luncheon, after which we wash it and go home, first carefully returning the fish to the stream from which they came. Such is a brief description of a fairly successful day spent ova hunting, the result being some three or four thousand eggs. There were often many blanks, however.

One November morning we started at four o'clock from Grange in Borrowdale to walk to Buttermere. It was cold, clear, and frosty, and "by the pale light of stars" we partially ascended the hill whose summit forms the famous Castle Crag, and then bearing to the right we skirted the fell known as Borrowdale Hause. By the time we had gone a couple of miles a snowstorm met us, and as we proceeded the drifts in some of the gullies we had to cross became uncomfortable. By the time we reached the head of Honister Pass the ground was deeply covered with snow, and when we arrived at the place where we had leave to fish we found the water to be so full of snow broth that the fish were not obtainable. Most of them had probably run down into the lake. We tried in vain, and at last gave up the attempt as hopeless, and, with all our paraphernalia of nets and cans, commenced the homeward journey. Up under the famous Honister Crag, with its snow wreaths and black jutting rocks, we passed, and at last reached once more the top of the famous pass. By this time, fortunately, the snow had ceased falling, and we risked the hill, and more than once got buried in a drift for thus defying the elements. However, home was safely reached at last, though minus any ova.

On another occasion we walked from our quarters at Grange to the head of Ulleswater, crossing Watendlath Fell, and over the three bridges of Thirlmere—now, alas! no more--up over a spur of the great Helvellyn, and down the Glenridding Vale into Patterdale, and, by the kind permission of the proprietor, fished some water there, stayed the night and tried next morning, but with very poor success, for the fish were not up from the lake.

After luncheon, we started on the return journey, and found some beautifully curious snow caverns in crossing the mountains, where a stream had been snowed up, and the water flowing underneath had hollowed out the frozen mass. It was getting dark as we crossed the Watendlath Bridge, and the waters of the celebrated tarn looked black and sombre in the dim twilight, but, knowing the danger of being benighted on the hills in weather that did not look the best, we hurried on until we reached the road in Borrowdale. These days were most enjoyable occasions in spite of the weather, which was often very wild and stormy.

Now, however, with a well ordered set of ponds and a goodly stock of breeding fish the matter is a very different one. A net specially constructed is run through one of the ponds, and brings to the bank at one haul several hundred fine large female fish, which are at once sorted. The ripe ones are picked out by an expert at a glance, and placed in tanks close by, while the unripe ones are thrown back again or put into a spare pond, as may be deemed most desirable at the time. Having thus separated the spawners, a large number of males are similarly secured from another pond, and the choicest fish are selected and also placed in readiness in tanks.. A number of spawning dishes are then brought out, a table on which to place them, and a few clean rough towels. The dishes are thoroughly dried, and a number of female fish are thrown into a net, from which the operator takes them one by one. He holds each fish successively, vent downwards, over the dish, with the left hand grasping just above the tail, while the right seizes the head. A very gentle pressure applied with the thumb of the right hand upon the belly of the fish causes the eggs to flow in an unbroken stream into the dish, till by the skilful movement of the thumb, the whole of them are expelled. A novice invariably makes great bungling at this work;

L

but "practice makes perfect," and the writer has spawned and impregnated the ova from a hundred and forty trout in the comparatively short space of about an hour. This, taking into account the changing of the dishes, and other little delays, is pretty quick work, yet the whole is done without any hurry or excitement; indeed, nothing is more undesirable than to perform the work too hurriedly, although, at the same time, expedition is of vital importance.

Having spawned a number of females into one dish, a batch of males is handed in a net by an attendant, and a couple of good milters taken, which is quite sufficient. These are somewhat similarly treated, the milt being expelled upon the eggs, and both eggs and milt are gently mixed by an undulating motion of the dish, aided slightly by the hand. Water is then added, and the mixing process is repeated, and the dish allowed to stand until the eggs have separated. At first they cling together, adhering not only to each other but to the dish itself, sometimes for a few minutes and occasionally for an hour or more, according to the temperature. On separating they must be well and carefully washed until all effete milt is cleared away. They are then placed upon the grilles in the hatching boxes, a number sufficient to fill each grille being poured out of a measure, which holds just the right quantity. They are poured from the measure so as to form several little heaps on each grille, and these heaps are roughly spread by means of a feather, care being taken that it does not touch the ova. The grille is then raised slightly, both hands being used in the operation. Then, by means of a slight shake, giving the grille a motion soon acquired by practice, the eggs are made to arrange themselves in rows very prettily, and they can be accurately and quickly counted.

In the old days there was often a difficulty in getting milters, and even to the present time collectors of ova from wild fish have often to be content with such milters as they can get. It will be seen what a great advantage it is to have a large number of well-bred and selected male fish in a pond by themselves, so that a haul of the net brings to bank several hundreds of them. Out of this number the best fish can be selected and sent up to the spawning house, where operations are being carried on. From a

AT SPAWNING TIME, SOLWAY FISHERY.

number of good fish a couple of milters are selected—sometimes one only is used, but as there are eggs from several females in the dish there is no harm in using more than one' male, should it be deemed desirable to do so. Sometimes the milt may not be quite ripe, or may be spent, or stale, and in such a case the ova would probably not be impregnated. By taking a second or even a third male under such circumstances good eggs may often be saved. I have seen the time when good milt was difficult to obtain late in the season, even from the ponds of a well-ordered fish farm.

These matters must be learned by practice and experience. It is as needful now-a-days for a would-be fish culturist first to go as a pupil to some first-class fish farm, as it is for anyone wishing to be thoroughly well up in agricultural pursuits to go and learn farming. There is much that cannot by any means be learned from books, and there is much also that one who is not already skilled in the work must be taught by an expert, not only practically but theoretically. A thorough knowledge of the development of the embryo under varied circumstances and surroundings is very essential, and can only be acquired by a good deal of thoroughly practical work with the microscope.

When viewed through a high power immediately upon extrusion from the fish, the milt is found to contain an enormous number of minute organisms called spermatozoa, which at first are very lively, and move about rapidly by means of flagella. Very soon their activity begins to cease, and in less than a couple of minutes they usually cease to move. In one part of the shell of the ovum is a minute aperture called the micropyle, and by means of this micropyle one of the spermatozoa is enabled to enter the egg and impregnation is effected. The reason for being expeditious in the work will be at once apparent when these facts are considered.

During the first few weeks of their existence the eggs are very delicate, and a very slight shock or vibration will often so disturb their organism as to kill them, but strange to say, this does not apply to the first twenty-four hours or more after they are taken from the fish, during which time they will bear any reasonable amount of shaking, but are very sensitive to temperature. This

seems a wonderful provision of nature for enabling man to
collect eggs and carry them at once hundreds of miles by rail, &c.,
to the hatchery if needed. I have known them nearly forty-eight
hours on a journey without taking harm, but after the first twenty-
four hours the sooner they are laid down in the hatching apparatus
the better.

A fully developed unimpregnated ovum consists of a mass of
protoplasm, in which, a little to one side may be seen a small
clear nucleus or cell called the *germinal vesicle*, and which in its
turn contains a still smaller cell or nucleolus—the germinal spot.
Some hours after being taken from the fish and laid on the grilles,
the germinal vesicle in each egg may be seen on the portion of
the ovum which is uppermost, and should such an ovum be
fertilized by spermatozoa, great changes soon take place, the first
of which is that the cell consisting of the germinal vesicle and
germinal spot is split into two cells, each of which in its turn
forms two, and so on in geometrical progression, which simple
cleavage, known as segmentation, continues for some time. At the
end of this process the ovum is a mass of nucleated corpuscles
without cell walls, and has reached the second or *morula* stage, so
called from its likeness to a mulberry. The cells on the surface
of this mulberry mass gradually become elongated or column
shaped, ending in long threads of protoplasm called cilia, by
means of which it can not only move through the fluid but
produce currents in it in its immediate neighbourhood. The
ovum has now reached the third or *planula* stage. At this point
a groove appears, down the centre of which a white streak, soon
taking the form of a ridge, may be seen, which is the *chorda
dorsalis* or *noto-chord*, which becomes enclosed by the wall of the
rest of the organism on either side growing over it. If a section
be made through this embryo, three layers—an outer, a middle,
and an inner—may be easily distinguished by the aid of a
microscope. From the outer layer, the skin, brain, and spinal
column, are developed; from the inner, the lining of the
alimentary canal with its appendages; whilst the middle layer
forms the rest, which is by far the greatest part of the organism.

When an ovum has not been impregnated it remains
unchanged, except that the germinal vesicle is differentiated as

in the impregnated eggs, but no further development taking place it simply remains, presenting the appearance of a globular shaped body, and in about three weeks or a little more, according to temperature hastening or retarding the development of the fertilized eggs, these unfertilized ova may be picked out. They are commonly called "blind" eggs because they never show the eye spots, which are, later on, such a prominent feature amongst the good eggs.

The so-called "dry method," by which trout eggs are now taken was discovered in Russia, by M. Vrasski. Experiments had also been tried in France to test the vitality of milt, and these had led to the discovery that the spermatozoa could be kept alive much longer out of water than in it. If the air be kept from it, as for instance when taken direct from the glands into a tube and hermetically sealed, it will sometimes keep good for a considerable length of time, and I have sent it by post to a friend, who found it to be quite fit for use at the end of twenty-four hours or more.

By this "dry method" of impregnation a much larger number of eggs is rendered fertile than formerly, when a spawning dish half full of water was used. In those days we used to think seventy-five per cent. and less a very good result; now we get ninety-five per cent. Sometimes, all the eggs are impregnated, but usually a few escape, and these are afterwards picked out, and will be alluded to in another chapter. Talking of percentages reminds me of a very irrelevant question a fish culturist was once asked by a juror at an exhibition. Instead of confining his queries to the subject of the exhibits, the juror asked what was the greatest percentage of alevins he had ever obtained from the ova laid down in his hatchery? He knowingly answered, "a hundred and one per cent.," which for a moment puzzled the judge, until he remembered that one egg sometimes yielded two or even three fish. Do not be disappointed if you do not always get a good result at first. There are many things to learn in connection with this important branch of the work. Sometimes the ova may not have been properly developed in the ovary; sometimes the constitutional weakness of the fish may be to blame; sometimes the eggs may have absorbed water; or even, owing it may be to a defective *micropyle*, the spermatozoa fail to enter. It has been suggested that in the case of using large males for the ova of small females, that the spermatozoa are too large to enter the *micropyles* of the ova of small fish. This may or may not be so, but the greater probability is that the diversity in the ages affects the fertilizing power, or too often the health of the embryos, or the future offspring. The age of the milters is I am convinced an important consideration, and I prefer a comparatively young fish to an old one—indeed, I do not keep male fish more than seven years as a rule; better knock them on the head and market them.

I must here, however, add a word of caution against using very young fish as milters. Two-year-old males will yield milt, and occasionally yearlings do so, but they should not be used. The late Dr. Francis Day records a series of experiments ("British and Irish Salmonidæ," p. 26), which go far to show that both ova and milt from young fish are not of good quality. Fish culturists certainly know this to be the case as regards ova, and they know also, that the males mature earlier than the females. Therefore, the probabilities are, as regards old breeders, that by using selected males that are younger than the females excellent results will follow. A great deal depends upon the selection of the fittest, and the more we come to understand this, the greater the measure of success that attends our efforts.

Formerly, the condition or age of a milter was never thought of. Any fish that came to hand was taken, and probably even now little attention is paid to the matter on some of our rivers when fish are scarce. It is important, however, in the case of salmon and other wild fish, to select suitable males, as it is amongst fish on a farm, and must tell on the future crop of those in a river.

The spawning of the first fish is to the novice usually a time of much interest, and often of some excitement. To tell exactly when the fish are ripe, is a point about which many who have consulted me evidently felt a little anxiety. There is not much wonder at this, and experience must be bought in one way or another with regard to dealing with fish at spawning time. A beginner often gets on fairly well up to some point, where a mistake is made which upsets the work a good deal, and for the benefit of those who are about to commence for the first time, and with a view to preventing blunders of such a kind, I will briefly give some simple directions.

There are so many ways of capturing the fish that I will not enter much into that subject, beyond saying that we use the most convenient and effective nets that ingenuity can devise, for taking the fish in the quickest and simplest way without hurting them. Much must depend upon the nature of the stream from which they are taken ; in some places a simple bag-net will do, in others large landing nets are very effective, and in some a carefully constructed trap-net in which the fish is caught in a bag the moment he strikes is a good net to use. The fish are not being captured by way of sport, but for the most excellent of all reasons, that we may care for the eggs, amongst which there is such great destruction when left to nature.

The fish then are to be handled very gently and taken great care of, and after we have "stripped" them of their eggs they will be very carefully returned to the water from which they came. It is the greatest mistake to suppose that any harm is done to them by taking their eggs. I never yet saw a fish killed by being spawned.

There is danger in leaving them too long in tubs or cans, however ; they may die or jump out, and they should, therefore, be detained as short a time as possible, and should be carefully watched while they are waiting. An excellent safeguard is to stretch a piece of fine netting over the tub and turn a current of water through it, which will keep the fish in a lively condition for an hour or two. I have used perforated cans or cages sunk in the stream, and fish may be safely left in these for several hours if necessary. If not crowded they may be left over a night when circumstances really demand it, but I never keep them so long if I can avoid it, as they rub the slime off their bodies and require to be dipped before being set at liberty.

Having provided the necessary nets and a tub or tubs, the fish caught should be sorted. To do this place the bag of one of the landing-nets in a tub so that the fish can be placed in it and kept there. A net such as the one described in my chapter on stocking, and which is used for reviving the fish before or after a salt bath, is very useful, and may be worked in the stream. Put the ripe spawners or females in one net, and the males or milters into another so that they can be got at at once without any uncertainty. Some towels and a few spawning dishes should be at hand. Milk bowls will be found to answer the purpose very well. When the spawning ground is some distance from home a few enamelled metal basins are very convenient, being light and easily carried, and a can will be required for taking home the ova. Something in the shape of an ordinary milk can with a lid will be found useful for this purpose. A bedroom water jug is an article I have often used, and seems to be about as good as anything. Whatever is used, take care that it is perfectly clean.

For collecting salmon and other ova, when large quantities are likely to be taken, I use carboys, and find them very convenient. They should be filled with water, and the eggs poured in from a jug or other vessel. The specific gravity of the eggs being rather greater than that of water they go at once to the bottom, displacing some water in the process. When about two-thirds full of ova a little of it should be poured off, and the bottle sent forward to the hatchery with as little delay as practicable. On arrival there the eggs may be poured out into bowls, which should be held immediately under the nozzle of the carboy during the operation. Should it be desired to extract them without any jar or concussion it is quite easily done as follows :—Place a bung or stopper of any kind in the neck of the carboy for a moment, and invert it over a tank with the neck submerged, withdraw the stopper and the eggs will quietly gravitate to the bottom of the tank.

When the fish have been duly sorted and are at hand in their various receptacles the work is easy enough. Some operators kneel down, and for beginners this is perhaps the best plan, though it is never done at the Solway Fishery. Some fish are almost sure to slip through the fingers of a novice at first, and they are not likely to be so much injured as they would be by a fall from the hands of a person standing. Even an old hand will let a fish slip occasionally, but the occurrence is a rare one, and the chances are that he so balances it, or dexterously controls its movements whilst falling as to send it into a tank or tub of water, from which it is soon removed unhurt. But the novice is as likely to send it into the spawning dish as anywhere, and a few struggles there will send the eggs flying in all directions.

The cleanliness of all apparatus used is of vital importance, and much of the success of the undertaking depends upon it. A clean and perfectly dry spawning dish and clean hands are most desirable. Take care that the fish are clean also; they will not be so if the net be placed on the ground. By retaining them in nets which can be lifted bodily they are in excellent condition for handling, and the water is shaken off them before the handling commences. Notwithstanding all precautions, it is impossible to prevent a drop or two getting into the spawning dish occasionally, but avoid it as much as possible, as the dryer the eggs are kept, until the milt has been added to them, the better. A beginner is anxious to know how long the eggs should remain in the dish with the milt. This depends chiefly on temperature; when they cease to adhere to each other, or to the dish, they may be washed, and this may be in ten minutes. Do not on any account wash them too soon. It is better to leave them for a couple of hours than to disturb them before they are ready. A little experience here is better than a good deal of theory.

If the day be warm and the temperature of the water 50° Fahrenheit there will not be a long time to wait, but should it be freezing hard and the water at 33° or 34°, then the time will be much longer. Take care not to drown the eggs with milt. They will take no harm standing in the dish for twelve hours after being washed, but they may suffer if the milt be left on them. I have frequently taken ova in the afternoon of one day, washed it, and allowed it to stand all night in bowls; and taken it over to the hatchery next morning to be laid on the grilles. Milters should always be carefully selected, and no fish that are in any way deformed should be used for breeding purposes. Many deformities are more or less hereditary, as has been proved in the case of other animals, and it is better to avoid them. Although experiments have been tried with the result that ova from deformed fish have apparently produced well-formed healthy fish, yet the reverse has also been ascertained to be the case. The result of "in-breeding," too, is most disastrous, and change of blood and judicious crossing of races and varieties is an absolute necessity if successful fish farming is to be carried on. An accepted theory amongst biologists is that though the reproductive cells as a rule vary considerably in size between one genus and another, it is less marked in the species, and disappears (especially in the male element) in the varieties.

An expert can often tell at a glance whether a fish is ripe or not. There are signs which it is difficult to describe, such as the appearance of the organs, the looseness of the ova in the fish, and a general appearance of ripeness, which can only be detected by long practice. No force should be used in expelling the eggs. If they do not flow freely on the hand being very gently passed over the belly of the fish they are not ripe. Of course a slight pressure is necessary, but far more depends upon proper handling. The hand or thumb should be kept behind the eggs, and not allowed to run over or in front of them. In taking the milt a knowledge of the situation of the glands is useful. They are much lower down than the ovaries of the female, and instead of using the hand, the milt may be easily obtained by an adroit use of the thumb and forefinger. The weather has much influence on the spawning of fish. A mild day, preceded by a warm rain, will make the fish lively, and they will spawn freely ; but a hard frost, or snow in the water, will retard the shedding of the ova, and consequently, on a fish farm, preparations are often made accordingly. We know, as a rule, before commencing, how the fish will yield their ova.

When the eggs have had the milt on them long enough, they are easily washed in the spawning dishes, by pouring off a part of the water, adding more that is fresh and clean, and continuing the operation some six or seven times, when the eggs should be clean. Be quite sure that they are so, and should there be any doubt about it give them another wash. At the Solway Fishery, where eggs are dealt with in large quantities, the contents of a dozen or a score of spawning dishes are emptied into a well-charred wooden tank, through which a current of water flows, and are left there for half-an-hour or more, when they are found to be perfectly clean. (Fig. 13).

One of these tanks will wash a hundred thousand eggs at a

time, and the saving in labour is thus considerable. When small quantities are to be washed they are done in buckets which are kept scrupulously clean, and are used for no other purpose.

It is quite easy, as a rule, to distinguish the sexes at spawning time. The males are in their best livery, and are much flatter and thinner in the body than the females, which are full and rounded. With a little practice they can be distinguished at a glance in most cases. There are a few exceptions, some specimens being a little deceptive in appearance. I have met with female fish yielding ova that have otherwise had all the outward characteristics of the male ; and I have met with male fish which at sight I have mistaken for females, but such cases are rare. The same freak of nature occurs occasionally amongst birds, an undoubted egg-producing female being found attired in the full plumage of a male, and *vice versa*. One of the most remarkable instances on record is the case of a hermaphrodite trout, which was discovered by Mr. Thomas Andrews, of Guildford, and presented to the Museum of the Royal College of Surgeons. The fish produced both ova and milt, and the eggs hatched.

Fig. 13.

CHAPTER VII.

Everything in perfect working order—Everything well seasoned—Preparing the grilles—Laying down ova—Picking—Beware of fungus—Sediment—Effects of concussion—Washing eggs—The eye spots—Embryo as seen through the microscope —The eggs commence hatching.

FOR the incubation of ova it is very important that everything should be in perfect working order. No work should be delayed in the preparation of the hatching house for its important duties. All painting and varnishing should be done three months beforehand, so that it may be not only dry but thoroughly hardened. Having completed all necessary work and alterations in good time, the water should be turned on a month before spawning time. All that will then be needful will be to cleanse the hatching boxes and aqueducts prior to the laying down of ova. However clean and pure the water may be this operation is desirable, for some sediment will probably have accumulated, as the filters have not been working, it being needless to put in the filter cloths or screens until a day or so before the laying down of ova. They should be all ready in good time so that nothing is required but to slip them into their places. The filter tanks should be cleaned as well as the hatching boxes, and before the water is turned on again the filter screens should be in their grooves, the lids should be on the hatching boxes, and all should be in thorough working order. It is a great comfort to have all in readiness beforehand, and is a necessity where good work is to be done. There are many little items that require attention, and it is only a case of exercising a little forethought, and not leaving everything until the hatching season.

Where any fresh concrete work comes in contact with the water I prefer running the latter over it for three months. Cement is partially soluble, and in some waters more than in others. The soluble portions are injurious to trout ova, but after being in use for some time the whole becomes thoroughly seasoned, and is then quite harmless. I am alluding now to such works as settling tanks and aqueducts, for which concrete will often be found very useful. It is a very good plan, after the water has been running over new works for a few days, to turn it off and give the whole a good scour out, and after a few more days repeat the process. After this, turn on the water for good, and do not on any account scour out again. Nature usually provides a coating for concrete work in the shape of minute vegetable growths, and these should not be thoughtlessly removed. The same applies to concrete ponds, with the exception that sometimes confervoid and other growths begin to threaten trouble. They may then be gently raked off or otherwise partially removed, but without any scraping.

The well-known proverb, "Do nothing rashly," certainly applies in fish culture, and it is well to observe it. It often happens that mistakes are seen when too late, which by a little forethought might have been avoided.

A day or so before spawning, a sufficient number of grilles should be placed in position for receiving the ova. They are usually stacked away in some convenient place, and are all ready for use. Dust them if they require it, and wash them well before placing them in the hatching boxes. It is well to keep them submerged in some tank for a few days, and finally to wash them under a tap. They are then ready for work and may be placed in position. On putting them into the hatching boxes, let one side go right to the bottom and shake them gently. This causes the glass tubes to fill with water. Then, without taking them out, lay them on their rests. They have a tendency to float at first; but this is easily counter-balanced by placing a piece of lead on each end of the grille. These weights may be fixed or loose, and I prefer them loose, and use neat little cubes of lead specially cast for the purpose. Small blocks of stone will do perfectly well, or even stones out of a brook.

When a box is commenced with it should always be filled

with grilles. Never half fill one, and lay down ova before putting in all the grilles. There would be a danger of disturbing the eggs in the lower half of the box, whilst placing grilles, etc., in the upper. Many persons would probably say that they could do it quite easily without doing the slightest harm, and I have no doubt they could—but there is no object to be gained, and harm may be done. It is one of those points the observance of which tends towards success, and I would not give much for the results from a hatchery where such rules were not strictly observed.

In laying down the eggs always commence at the lowest end of a set of boxes, and go on filling as the ova comes into the hatchery, until that set or series of boxes has received its full complement of ova. In a large hatchery it is sometimes desirable to have several sets of boxes filling at the same time. This does not matter. The object is that as the eggs are laid down, so shall they hatch in rotation, and it would be bad policy to arrange to have some hatching in one of the upper boxes, whilst others were being incubated in a box below.

There will probably be a few white eggs, and these should be picked out. This may be done either before they are laid down or after, or if the former course be taken both before and after, for some will probably turn white either whilst being laid down or very soon after. They are easily picked over whilst in a bowl or other vessel, as owing to a slight difference in the specific gravity, the white eggs rise to the top when slightly agitated, either by pouring in water or by means of the hand. The reason why some of them turn white is that they have absorbed water, which acts on the albumen of the egg, causing a white precipitate to form and this goes on more or less during the whole time of incubation. Where a good impregnation has been secured there will not be many white eggs. It is the badly impregnated and unimpregnated ova that turn white as soon as the water percolates the shell of the ovum. Some unimpregnated or " blind eggs " remain unchanged until all the others are hatched. These will do no harm, and may either be picked off any time at leisure, or left until all the hatching is completed, when they can be removed with the grilles.

A variety of implements have been used for the purpose of

picking ova. The one that I have found best, and have used for a quarter of a century, is very simple in its construction, and is easily handled after a little practice. It consists merely of thin brass or copper wire twisted into a loop of the proper size to bring out the eggs. It may, if desired, be fixed into a wooden handle.

It was invented by Mr. Seth Green, the veteran fish culturist of the United States. There are many other useful appliances, most of which I have tried, but have not yet met with any implement that will do better work, or that is so exceedingly simple in its construction. To use it, place the loop under the egg to be withdrawn and lift quickly. Some dexterity is needful, and is easily acquired. As the whole secret of using it lies in the pressure of the water keeping the egg in its place in the wire loop, hence the need for rapid action. A beginner should practise on some lots of newly-fertilized ova or on some which is well "eyed," as a slip or two then will not do any very material damage.

Fig. 14.

The other tools used for this purpose are tweezers, pliers, suction tubes, etc., and these are all found very useful by those who prefer them. A tube fitted with an india-rubber bulb does very good work, and one of my pupils won the day at egg-picking with one of these instruments against the wire loop-picker. It is not, however, the man or the tool that can pick the most eggs in a given time, but the one that can pick them without injury to the others. I have seen a clumsy prac¦titioner getting on very well as to quantity, but doing more harm to the rest of the eggs than any good he was doing by picking out those that required it. Egg-picking would probably be well done by those experts known as the "light-fingered gentry," could they be persuaded to earn an honest livelihood. In my hatchery it has been largely done by girls, and also by men and boys, but to whatever class of workers it is entrusted, they should fully under-stand the need for care and skill in not disturbing the good eggs.

It is necessary that all white eggs should be picked out, for if left too long in the boxes the fungus grows upon them, which also surrounds the good eggs lying in their immediate neighbourhood, and injures or destroys them also. In a well ordered hatchery this work should be done daily, even though there be but few eggs to pick. There will be more amongst some lots than others, but all the boxes should be carefully gone through. This fungoid growth on the dead 'eggs is usually known amongst fish culturists as "byssus," in contra-distinction from "fungus" *(Saprolegnia)*, which attacks both living fish and living eggs.

The "fungus" *(Saprolegnia)* will attack both living and dead fish, but the "byssus" *(Leptomitus clavatus*: STONE*)* only grows on the dead. When it occurs amongst eggs, as it most assuredly will if the white ones are not picked out, it grows rapidly and sends out innumerable filaments, which soon surround the good eggs within its reach and cause partial suffocation. Such an occurrence should never be allowed to take place, and is easily prevented by punctually removing the white eggs as soon as they become opaque. I have known "byssus" appear within twenty-four hours, which shows the necessity for prompt action. Should any of the surrounding eggs be taken within its grasp, they may not die immediately as a result, but the embryos are weakened and cannot be expected to make good fish. If white eggs, after being picked, be placed in a glass of sea water they will become clear again, and on being transferred to fresh water they again resume their opaque appearance, and the experiment may be several times repeated with the same eggs.

Sometimes healthy ova die and become opaque owing to being attacked by living enemies, which have got into the hatching boxes. It is in fact, exceedingly difficult at times to keep these out. They come in the shape of eggs deposited in the water, or in other stages, when they are very minute, and they soon grow bigger and cannot get out again. These will attack the ova and puncture the shell, which then allows the water to enter, and the precipitation of the albumen follows as a natural result. With good filters, cleanliness, and everything working efficiently, the loss from this cause ought to be very small indeed.

The "fungus" *(Saprolegnia)*, to which I have referred, sometimes attacks the eggs, and woe to them should this happen. It ought never to be allowed to do so, however, as it will not readily grow upon them of its own accord. They do not form one of its natural habitats, and should it make its appearance, depend upon it something is wrong somewhere. Three conditions are necessary for the avoidance of *Saprolegnia* :—(1) Darkness; (2) thorough cleanliness throughout; (3) and all wood to be carbonized.

Light is highly favourable to its growth, and darkness is unfavourable. Darkness is also good for the eggs, and light is the reverse; indeed, too much light is calculated to injure the embryos. Therefore, keep the boxes always covered. As regards cleanliness, it should not be necessary to say much, and yet I know it is sadly too often the case that for want of it the eggs suffer. The filters should be thoroughly effective, and every precaution taken to keep the hatching boxes perfectly clean during the period of incubation. Every scrap of woodwork below water-line should be charred, and all the joints and knots or any other places that look at all suspicious should be varnished also the first season. For the second season at least two coats should be applied, and each succeeding season afterwards one or two coats, as may be deemed desirable. Especially about any nail heads, and the corners and crevices of the grilles should the varnish be applied.

The inside of the boxes should never be too roughly scoured out after they have been varnished for the season. It must be remembered that every scratch or puncture which lays bare the wood, creates a suitable *nidus* for the germs of the fungus. I have seen cases in which such little regard has been paid to this that there has been little cause for wonder at an outbreak of the deadly pest. Should anything go wrong with the filters and the eggs get covered with sediment, there is at once a danger of fungus making its appearance. The two things often go together.

The eminent American fish culturist, Livingstone Stone, says :—" There is no word in the fish breeders' vocabulary that is so associated with loss and devastation as the word 'fungus.' There is nothing with which he has to deal that is so insidious and deadly. This silent invisible foe is sure to come if any door is left open for its entrance. It often fastens its irrevocable grasp on the eggs, without giving any sign of its approach. Once present in the water it spreads over everything. It cannot be removed. It never lets go its hold. It is fatal in its effects."

M

Take care to keep everything very clean, therefore; keep the boxes covered, and have all the woodwork well varnished, and there is little to fear from this enemy.

Sediment is also a danger which should be carefully guarded against. Unless great care be taken it is sure to come, and to settle more or less upon the ova. Although I have known cases in which sediment has apparently had no very prejudicial effect upon ova, yet other cases have come under my notice in which very serious harm has accrued. At some periods of their existence eggs would not be so much affected by it as at others, and the sediment itself varies very much according to circumstances. In some waters it would be almost entirely mineral, whilst in others it would be chiefly of a vegetable or organic nature. Or it might consist of both mineral and vegetable matter. Some deposits are directly poisonous, as for instance in cases where copper or lead are present. Others again are highly injurious, as for instance iron or lime, although I have seen good work done in water containing both. But assuming that a sediment could deposit on the eggs, of such a nature as to be quite innocuous in itself, the result must tend towards suffocation. This may be only partial but, nevertheless, it is sufficient to arrest the development of the embryo, and as a consequence deformities are produced. It is well known that the ova of salmonoids require a good supply of oxygen. They likewise give off carbon, in the form of CO_2, which requires to be carried away from the eggs as produced, and when covered with sediment this healthy change of condition is not maintained, and should it remain long enough many of the eggs will be poisoned. Although they may be not absolutely killed at the time, the result will probably be weakly or deformed fish.

When eggs are packed in moss and sent long distances cripples usually occur in excess amongst the fry produced. This has been notably the case amongst eggs imported from abroad. It is not at all surprising that this should be so when the development of the embryo is considered. The pressure, slight as it may be, of the layers of moss upon the eggs, is quite enough to cause the many curved spines and other deformities, that occur under such conditions. Concussion has also been found to produce a similar effect, and often to cause the death of the embryo.

Dr. Day records (British and Irish *Salmonidæ*, page 41) an interesting experiment which he tried "in order to ascertain the effects of direct concussion on ova. A number were dropped from various heights, or direct into the water, when it was found that in those in which this was tried within twenty-two days after being obtained from the fish, none lived over eight or nine days." During the first stage of the ovum after impregnation it will bear a reasonable amount of concussion, but after a period of some_ thirty-six hours, varying according to temperature, a very slight amount will suffice to destroy the vitality of the embryo. It naturally follows that in cases where the latter is not killed it receives serious injury which will tell upon its after-life.

I remember a tray of eggs getting accidentally shaken on one occasion when incubation had proceeded for nine days, and although the accident at the time seemed trifling and the concussion was slight, yet within a few days over twenty-five per cent. of those eggs were picked off dead. After the eggs are well eyed concussion does not seem to kill them, as in their earlier stages, but it produces injuries which often seriously affect the after-life of the fish.

The cure for sediment is an efficient filtering apparatus. Should it, however, from any cause settle upon the eggs, they may be washed after they have reached a certain stage, but care should be exercised in the operation. It is not safe to attempt it much before the eye spots appear, or when the eggs are about half incubated. It may be done earlier by an expert, but the safest plan is to leave it until the eggs will bear moving. The water may then be lowered and a watering-can used, or they may be gently feathered over without lowering the water.

The best plan of all, however, if there be much sediment, is to remove the eggs and wash them, and clean out the hatching boxes thoroughly. To do this it is desirable to have one hatching box at liberty, which should be made ready for eggs. Then the contents of another box may be lifted out a grille at a time with the eggs upon it. It is needful to have a movable tank or vessel capable of receiving a grille, and this being filled with water, one may be placed in it and reversed, and then another, and so on.

The tank used for washing the eggs when first taken from the fish will answer admirably (see Fig. 13, page 155) for this purpose, and with a good current of water flowing through it, twenty grilles or more may be reversed in succession and all the eggs washed at once. Much care is required in carrying out a piece of work of this kind. It should not be done thoughtlessly. Eyed ova are very easily suffocated, and this fact should be borne in mind at all times when it becomes necessary to deal with them in large masses, or when they happen to be placed temporarily in bowls or other vessels containing still water, as may be the case when packing for a journey.

Should any eggs adhere to the grilles, wash them off gently. The dirt will by this time have mostly separated from the eggs, which may be washed in the usual way as recommended at spawning time, great care being exercised in the operation. When as clean as they can be made, replace them on clean grilles in the newly-cleaned hatching-box, and repeat the process with another lot, and so on through the hatchery. With regard to the washing of younger eggs, I would advise the reader, if disposed to try it, to experiment on a few first, and note the result. More practical information will be procured that way than from any books. The best way, however, is to guard against sediment, and there are few places where it cannot be kept out of a hatchery. The filter screens should be cleaned twice daily, or oftener, if required. To do this, take them out one at a time, and wash them under a tap. A brush will be found useful for this purpose.

During each day of the period of incubation there will be a few white eggs to pick out which should be duly attended to. After a washing process there will probably be a great many, or what will appear to be a great many in comparison with other days. There is not necessarily any cause for alarm at this, as the agitation will cause the water to percolate the shells of a large proportion of the blind or unimpregnated eggs, which will thereupon turn opaquely white. There will be less picking to do afterwards—at least there ought to be—and should this not prove to be the case, then harm has been done to the eggs during the washing operation.

The first appearance of the eye spots is always an interesting sight, and especially to those who have only just commenced practical work. In water which stands at about 40° to 43° Fahrenheit ova will hatch in about ninety to ninety-six days, and in about forty-two days or a little more the eye spots will begin to appear. The time varies a little in the different species of *Salmonidæ*, and in the case of hybrids is often shorter than in that of pure bred fish. The higher the temperature of the water the sooner will the eggs hatch, and the colder it is the more they will be retarded. Salmon ova have been hatched in thirty days, and they have, in very cold water, been as long as one hundred and sixty days.

In the year 1883, by using ice freely to lower the temperature of the water, I kept some back for a hundred and thirty-two days, and by so doing was enabled to have them hatching in my boxes at the Fisheries Exhibition at South Kensington on May 12th. They were the only British ova of *Salmonidæ* in the exhibition. They were sent from the Solway Fishery in water in bottles, and travelled perfectly. In the warm London water they soon hatched, and a few days later I was much amused when explaining this to one of the spectators who came to see them. He gravely said, " Haw, can't you wepeat the pwocess?" Another seriously asked—" How long a salmon had to sit on its eggs before they were hatched?" Surely there is room for some instruction in fish culture.

In order to keep the ova in a healthy condition it is necessary to run a good current of water over them. The quantity needful varies according to the nature and quality of the water. I have seen apparently good work done with a small flow; indeed I have myself, in early days, hatched salmon under a tap, from which a mere trickle of water came. It is quite possible to hatch ova without running water at all, by simply changing it twice daily; indeed, I may go even further, and say that they can almost be hatched without water, for I have kept them for many weeks and hatched them in damp moss. There are two points to observe in carrying out such an experiment—*viz.*, that there is free access for oxygen to the ova, and a means of exit for the carbonic acid exhumed; in other words, good ventilation. But to produce really good fish a copious supply of good water must be at hand. In the main hatchery at the Solway Fishery there is a supply of over two hundred thousand gallons daily, and a similar supply in an accessory hatchery, and in addition to this there are several excellent springs on the estate that have not as yet been used at all.

I was obliged to sacrifice the Troutdale hatchery and ponds in Cumberland owing to the shortness of the water supply, and would warn others against falling into the same trap. Fish culture in those days was on a much smaller scale than it is now, and the prices of ova and fish much higher. Without an efficient water supply cultivation on anything like a paying scale is quite hopeless. There should at all times be enough, and plenty to spare, so that in any emergency there is an ample supply to fall back upon. Through a box containing twenty thousand ova, from five to ten gallons a minute should be used, and this will do for a set of five or six of such boxes. The quantity of water is increased as the hatching time approaches, and after hatching, fifteen gallons a minute are sometimes run through each box.

I have already briefly traced the development of the embryo up to the formation of the chorda dorsalis or notochord. When about half incubated the two eye spots will become clearly visible, and as soon as they are fairly well developed the ova will bear handling, and an egg at this stage is a most interesting object when viewed under the microscope. The circulation of the blood can be distinctly seen as it courses along the chief arteries and some of the veins, in its passage to and from the heart, the pulsation of which may also be detected. The future skeleton of the fish may be traced, as well as the muscular fibre and tissues of the body, and from this day forward the whole presents an increasingly interesting object, the developing progress of which may be watched daily. Possibly during the first examination the tail of the fish may be seen to be possessed of a free movement. If not, it soon will be, and the deepening of the colour of the blood will also be observed about the same time.

The accompanying figure represents the ovum of a trout in a forward state, as viewed under the microscope. Owing to the position in which it is placed only one eye (1) is in focus. The breathing apparatus or gills (2) are very distinctly seen, whilst the heart (7) may also be detected and the pulsations distinctly noted, although, being a little out of focus it does not show out as clearly in this view of the ovum as in some others. For the same reason the body of the fish is not discernible, but the veins and arteries of the umbilical sac are very distinct. 3 is the artery which conveys the blood to the body, whilst 4 is the vein by which the same blood returns to the heart. 5 represents the vein which collect the blood from the capillaries ·or smaller veins. The small dots in the veins represent the blood coursing along. The

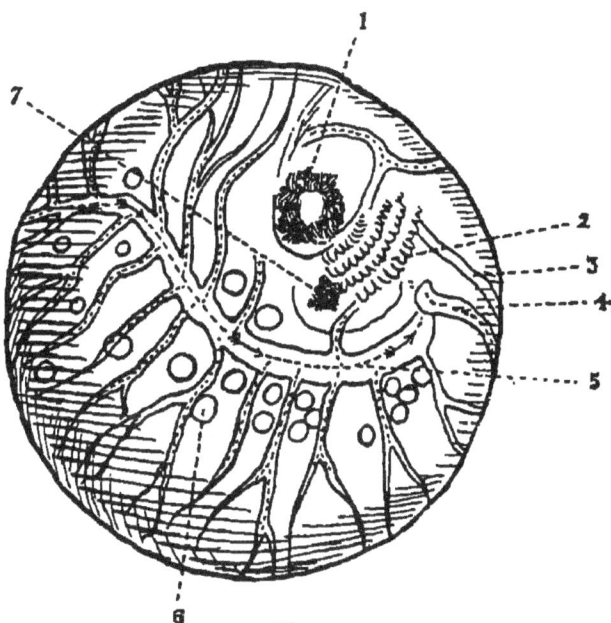

Fig. 15.

numerous rounded bodies (as No. 6) are the oil globules which supply nourishment.

The development goes on, and the embryo requires more oxygen. The water supply should be gradually increased, and daily attention given to the picking of the ova. When picking, keep the lids off the boxes as short a time as possible, and let the work be done in a modified light. A few may burst and some of the contents emerge, turning white immediately as the water comes in contact with the albumen. This is an excellent time for looking over the grilles and picking out any " blind " eggs

that may still remain, and also any containing "puny embryos."
The latter are now easily distinguished from the good eggs, and
may be detected at a glance even in a moderate light by the
difference in colour, being much lighter, and the eye spots being
very much smaller. The whole of the embryo is deficient both
in growth and stamina, and should such eggs succeed in hatching,
the fish will never live to grow up.

Some day, in looking over the eggs, a curious streak will be
seen amongst them, and on touching it with a feather, it will
become violently active, and will probably run away, carrying an
egg, or rather an egg-shell, in front of it, very much as a man
holds an umbrella in a gale of wind. It is the tail of a young fish,
which is the first to emerge, and is already endued with the power
that it is in future destined to exercise should it live :—*viz.*, that
of a propeller. Soon others will follow, and if the temperature of
the water be raised slightly a general hatching will take place, and
all the eggs of that particular lot may be hatched off at once.
Some unfortunates emerge head first, and unless helped a little,
by means of a feather or a camel's-hair pencil, they will probably
suffocate and die   There are more of these cases amongst some
lots of eggs than others, notably amongst char, and the manager
has sometimes saved the lives of several hundreds of these, by
giving them a little attention in the way already alluded to. It is
easier to help the little fish to escape from their prisons than to
pick them out dead afterwards, with all the accompanying *débris*
of burst yolk sacs, etc. It must be done one way or the other.

When the eggs are hatching or have hatched, a dipping tube
will be found a very useful instrument for removing little bits of
débris that are sure to occur in the hatching boxes. A plain
straight tube with a bulb is very useful for the purpose (see Fig.
17), or indeed even a plain straight one (Fig. 16) about three-
eights of an inch in diameter. Some persons prefer a bent tube
(as Fig. 18), and some use a tube that is provided with a cup and
indiarubber drum (as Fig. 19). This one is used by pressing the
drum slightly by means of the finger, before putting into the
water, and, on letting go, the object over which the tube is
placed is drawn into it, or is held at the mouth by atmospheric
pressure. The position of the hand (see 16) in the diagram will

show how to work any other tube. Simply place the forefinger
on the end of the tube as in the figure. Then place the other
end of the tube over the object to be withdrawn, and remove the
finger for an instant only. Take care to keep it on the end as the
tube is lifted out, and the object will be found inside it, be it an egg,
or an egg-shell, or a dead embryo. Anyone can soon get into the way
of using these tubes very readily. I have seen one used occasionally

Fig. 16.   Fig. 17.   Fig. 18.   Fig. 19.

that is fitted with an indiarubber bulb, and it also works very
well. A worker soon gets used to any particular form of instrument, and can naturally use that particular one best. The matter
is, therefore, a good deal one of choice.

During the period of incubation, as indeed afterwards, the
hatchery should be inspected early and late. It should be the
duty of some responsible person to look in last thing, before going
to bed, to see that the water is running all right, and then again
early next morning. A mere glance is sufficient. I once had a
man who carefully looked at each of the tail boxes to see that the
water was running out, and then went to the other end to see if it
were running in. He always did this, and I never checked him,
for he was thoroughly conscientious and trustworthy to a degree
in matters of this kind. When, however, the water is seen to be
flowing correctly at the outlets, there is no need to look further.
This ought to be a sufficient guarantee that all is right. I am
alluding now to the period of incubation only. After hatching, it
is a very different matter.

CHAPTER VIII.

HATCHING THE EGGS.

Glass grilles—Their cost—Their advantages—Cleaning the hatching boxes—The egg-shells—Artificial ova beds—Settling pond—Filtering bed—Wire grilles—Destruction of ova left to Nature—Advantage of artificial beds—Californian baskets—Repairing grilles—Overcrowding—Way of economising space—Compact storage box.

THERE are more ways than one of dealing with trout eggs. After a careful study of the various systems in use, and after weighing well the evidence, I give my verdict, and give it unhesitatingly in favour of the glass grille system. I have myself used grilles for more than twenty-five years, and have not yet met with anyone who can give one valid reason for their discontinuance. Of the arguments brought to bear, that which at first seemed the most plausible was the one of cost, but it would not stand scrutiny.

The chief item consists of the glass tubes or rods ; but seeing that, apart from actual breakages, glass will last for ages, there is little to allow for depreciation. As to breakages, we all know that glass will break, but with proper care there is but a small loss from this cause. I have carefully noted that which takes place in my own hatcheries, and find that out of every ten thousand glass tubes, about seventy-five per annum get broken, or considerably less than one per cent. As for the wooden parts, it is found in practice that they are less perishable than metal, for with care they last for twenty years, and are good at the end of that time. Grilles can be purchased ready-made for four shillings and sixpence each, and a grille holds from 3,500 to 5,000 ova. Taking 4,000 as an average, and allowing that a grille only lasts twenty years, then a grille incubates in the course of its existence 80,000 eggs, at a cost of less than three farthings per thousand, and the work, other things being equal, is done satisfactorily.

Glass gives off nothing, is easily kept clean, and does not require any varnish. Metals, such as are used, are at least partially soluble in water, and in some waters are very soluble indeed, and this renders necessary the use of varnish. A carefully prepared compound does very well, but some varnishes contain highly injurious ingredients, and here again is a considerable source of danger. Notwithstanding all drawbacks, however, I believe very good results have been obtained by using metallic trays.

I was one day discussing this matter with an American fish-culturist, who assured me that they had discarded grilles long ago, and he exhibited to me, and explained, some of the metallic trays and baskets that were in use in the United States and Canada, and said to be doing very good work there. In reply to the query, as to which system would show the heaviest death-rate, he at once replied that probably it might be a little greater on the metal than on the glass. His reply confirmed my convictions, and decided me all the more in favour of the glass system. I have since experimented myself, and have also had the benefit of the experience of others, and can only come to one conclusion. It is, that there exists at least one very great advantage in the grille system over every other, viz.—that stronger and better developed embryos are produced, and they have a far better chance of growing up and making good fish than any others.

The counting of the eggs, too, is so easily done on grilles, that in a large hatchery they are found not only a great convenience, but a great saving of labour. They also ensure correctness in estimating the number of ova laid down. The eggs arrange themselves so readily in rows that in a few minutes a large number of them may be accurately counted. For instance, a grille contains a hundred rows of eggs, and there are forty in a row. When multiplied this gives four thousand as the contents of the grille. A box contains ten grilles, which means forty thousand of those particular eggs. In another hatching box are eggs of a different species of fish, and of a different size. These can easily be counted in the same manner. The grille may be found to contain one hundred rows of eggs, and thirty-five eggs in a row, which gives three thousand five hundred eggs on a grille, and thirty-five thousand in a box.

When necessary, several kinds of eggs may be incubated in the same box, without any difficulty. Not only may each grille contain a different kind or class of egg, but in the case of small or experimental lots, which must occur in every hatchery, several groups of eggs may be placed quite separately on each grille, by simply leaving one row or space empty, to mark the division. This is found to be of the utmost convenience, as it often happens that these small batches of eggs occur unexpectedly, or on a very busy day when there is no time to provide a special corner for them. In a box set apart for such lots, they are simply laid down at once on the next grille, or part of a grille, in succession. When the eye spots appear, they can be moved to any other more suitable place, or packed and sent away, if required, right off the grille on which they have been incubated.

One great advantage which attaches to the grille system is, that all the eggs can at any time be seen at a glance, and the dead or white eggs picked off very readily, and should any sediment by any chance whatever have settled on them, during the period of incubation, it can be quite easily washed off, by lowering the water and using a watering can. This washing, if not done at too early a stage, will not hurt them, but will do them good, and it is just as well at the same time, to lift each grille with the eggs upon it, and clean out the hatching box, for with the purest water something will inevitably settle on the bottom during the two or three months occupied by incubation. Even the dust from the atmosphere, which is often considerable although invisible, settles down, and is continually being drawn in by the water, and deposited in the boxes and on the eggs. Of course, in a well-ordered hatchery there should be no sediment whatever, in the ordinary sense of the word ; and with proper appliances and precautions it may easily be kept out.

Having the hatching boxes thoroughly cleansed is an important piece of work, to be done before hatching, and it will be found a very good plan to have at least one spare hatching box. It can be made ready to receive the grilles from another box. This box may then be cleansed, and made ready for those from the next, and so on, right through the hatchery.

Thus the fish or "alevins" as they are called at this stage, have perfectly clean boxes in which to commence life, a most important matter considering the extremely delicate mechanism of the little creatures. More than this, it will be found, in practice, that on emerging from the egg they slip through between the glasses of the grille, and find their level on the bottom of the box, whereas the egg-shells remain where they were, with the exception of a few which are carried over the end of the grille by the "alevins," in their efforts to escape from their prisons. They are, therefore, at once separated, and all being hatched, the grille can be lifted out with the shells upon it and the fish will be found underneath it, packed together, somewhat reminding one of a swarm of bees. The eggs can all be hatched in a few hours, by regulating the temperature of the water, and in a large hatchery this is a matter of importance, and due provision should be made for it. All that is necessary is to raise the temperature a few degrees in the boxes in which the eggs are on the point of hatching. They can then be hatched off, and the shells lifted out with the otherwise empty grilles. There is no trouble from clogging of screens or anything else. The few shells which remain in the box may be left for the present, and will do no harm. Put on the lid and leave the whole affair till the next day.

The egg-shells remaining may then be cleared by drawing the plug behind the screen at the end of the hatching box, and so inducing an extra current. The water supply may or may not be stopped, according to the will of the operator, as he may find best. In either case the result will be the same—the water will be lowered and so lessened in bulk, the current will be increased, the "alevins" will remain packed, and the shells will be carried on to the screen. Care should be taken that the box is not so far drained that the fish are left uncovered even for a moment. After re-inserting the plug the water rises, and a bowl should be brought alongside and placed on a table which, when not in use, is kept under the hatching boxes

A syphon is then inserted, and by means of it the shells are easily drawn off the screen into the bowl, the overflow spout from the box being momentarily blocked by a piece of wood to stop the current. All this is done very quickly, and before the egg-shells in the bowl have settled they are gently poured off into a sieve which is attached to the table, and are there left, to be after-wards thrown away. It will usually be found that a few "alevins" are in the bowl. These will settle to the bottom, and it is quite easy to pour off the egg-shells without allowing them to pass over. They are probably weakly fish, and may be turned into some stream; or, if it be intended to keep them, put back into the box, taking care to put them in at the head, where the water enters.

One of the most simple and effective methods of stocking waters with trout is by planting eyed ova in artificial ova beds. By doing this, when the eggs are almost on the point of hatching, the heavy losses which occur in nature previous to that period are avoided. The beds themselves also, being made artificially, can be placed in the most convenient situations, where they will be entirely beyond the reach of floods. The water should be led into them through a pipe, or by an open spout. I have found an ordinary two-inch draining tile quite large enough, and as a few of these are usually easily obtainable anywhere, they are about as handy as anything that can be procured. A piece of common wooden spouting, if at hand, also does very well. It simply requires to be covered by a board, or a few slates, to keep out the dust, leaves, etc., and the whole may be protected from frost by a heap of straw, bracken, or heather. The hatching bed is not made in the stream itself, but somewhere near it, where it and its water supply will be absolutely under control. Under such con-ditions it ought to do very successful work, and the plan of hatching ova in such beds is of infinitely greater use, in many cases, than the turning out of fry, and is done at less than one fourth of the cost.

A bed may be simply dug out, and if this course be adopted, it is really nothing more than a ditch, which can be carefully guarded from birds, etc., when the eggs are laid down in it. It may be from one to two feet in width, and the amount of fall should be such, that at regular intervals, varying from six to ten feet, barriers may be placed, over which the water ripples from one section of the bed to another.   An excellent material for these barriers will be found in ordinary roofing slates.   I have used them with success for many years, and they are easily stuck into the bottom, lowered, or raised, or manipulated in any way that may be desired, and a lot of them should always be available where fish culture is carried on.   They are the cleanest and simplest articles to make use of for such purposes that I have ever met with.   They are cheap, easily cut to any shape, they can be laced together if needful by drilling holes through them, and, apart from accidents, are almost imperishable.

The depth of water should be about four inches, and the bottom should consist of clean gravel, the grains of which are principally about the size of peas.   Amongst this gravel the eggs are to be sown, and a section ten feet long and one foot wide will hold easily about fifteen thousand ova.   By making the width eighteen inches, the quantity may be increased to twenty thousand or more.   But with such an exceedingly simple contrivance there is no need to crowd the ova, and it is far better to err on the side of having too much space than too little.   The water, after doing duty in the hatching bed, passes on and finally re-enters the stream from which it came.   Should the width of the bed be increased, it should be borne in mind that a greater water supply is needful.   If two-inch tiles be used for a bed one foot wide, then a double set of them will be required for a bed one-and-a-half feet to two feet wide ; or a single course of three-inch tiles will answer the same purpose.

The ova may be laid on the gravel or mixed with it.   This is a question about which there is a difference of opinion.   When we consider for a few moments the requirements of the ova, and examine carefully the state of things in a natural trout bed, we shall very soon be in a position to judge which is best.   I have seen some thousands of trout nests, and I have invariably observed, that where the eggs are deposited naturally, water is found welling up through the gravel in which they lie buried—and often buried deeply.   Thus, the necessary conditions for their well-being are provided by nature.   The same conditions must be found in artificial beds.   The eggs must have the benefit of the continual change of water that has been provided, and if the current be all on the surface, it is clear that it would not be advisable to bury the eggs deep in the gravel. The best plan, in such a situation, is to place them on the surface, slightly mixed with the gravel, which prevents them from touching one another. When it is remembered that every egg gives off carbonic acid, the advantage of this will be apparent.

Some years ago I invented an artificial hatching bed, which has now been thoroughly tested and found to work exceedingly well. It is made of wood, has a lid, and is on the hatching-box principle. In size, it has been found convenient to make these beds twelve feet long, nine inches wide inside, and five or six inches deep. Along the entire length of the box, except for about two inches where the water enters, is fixed a false bottom of perforated zinc. The inflowing water is conducted under this perforated bottom, by means of a water-board placed a couple of inches from the end where the water supply enters. The perforated bottom is placed about an inch above the real bottom, and the water rises up through it. It is coated with asphaltum varnish, and covered with gravel, and upon this gravel the eggs are deposited. At the lower end of the box is the outlet for the escape of the water. This is formed by simply cutting away the upper portion of the box end, say for a couple of inches down. It will be clearly seen from this description that the water on entering the box at one end is conducted direct to the bottom, but passes out at the top, so that it must rise up through the perforated metal and the gravel and eggs upon it.

Gravel is beneficial in several ways. It tends to prevent over-crowding of the eggs, and it keeps them from coming in contact with the metal. It also provides a more natural bed for the "alevins," and as we are now endeavouring to treat them as nearly according to nature as possible, it is best that it should be so. I have always liked to associate young trout with a clean gravelly bottom as far as practicable, but in a large hatchery, where quantities arc dealt with, it is much better not to use it at all. The fish do better and are kept cleaner when they are the sole occupants of the rearing box. But in an artificial bed the case is very different. In the rearing box in the hatchery the fish are obliged to remain, whether they like it or not; but from the artificial bed they can escape at any moment that they may be seized with a desire to go. Indeed, perfect freedom is the secret of the whole thing. The little fellows have no recollection of having been shovelled about as eggs, and counted, and incubated on glass grilles. They find themselves in a very natural and tidy-looking place, and are content with their surroundings until a desire seizes them to roam, and when this occurs they must find themselves at perfect liberty to do so.

In addition to the dug-out system and the wooden one, both of which possess the advantage of being ready at very short notice, excellent work of a more permanent nature may be done by using concrete or brickwork. In this case, however, the work should be done in June or July, and the water turned on a few months so as to have everything well seasoned before hatching time. Cement is partially soluble in water, but this only applies to the time when water is first run over it. The soluble portions are soon taken up, and the face of the concrete becomes coated with a vegetable growth; after this it does no harm. The great advantages of these hatching beds are that they can be made available for use at once; that when once started no one has to attend to them; that they cost very little and do a great deal; they can be fitted up in a day, or a couple at most, on any stream, and the results are certain if good eggs be laid down. It should always be borne in mind that both ova and "alevins" should be kept well shaded from light. A box with a lid on is good, but the bed made in an open ditch must also be covered. A few old boards will come in very useful here. I have found larch slabs battened together do admirably, look neat and rustic, and last for years.

Only one thing more is needful for the successful working of an artificial ova bed, and that is an unfailing supply of water. I do not mean so much in the sense of quantity as in certainty. There must be no doubt as to whether the supply will continue. No choking of the screen or entrance of a frog into a tile must be allowed to cause a stoppage of the works. These things are easily guarded against, and proper precautions must be taken from the first to prevent such calamities. A very simple way of dealing with the water at the intake is to have a sufficiently large screen

N

over the pipe, which must be fitted into a box. To do this in the readiest way, take any old box and cut a round hole in one side of it to fit on to the end of the pipe. The opposite side of the box should be almost entirely cut away, or it may be knocked out altogether and replaced by perforated zinc. Should it be found needful a finer screen can be fixed inside the box, which itself is to be firmly embedded in the side of the stream from which the water is taken. In some cases it may be desirable to put on a coarser screen outside, and this is easily done by nailing four strips of wood on to the box front, and covering again with coarsely perforated zinc. Of course, where a properly constructed box is used, grooves can be sawn in the wood and three or four screens made to slide into their places, and these can be cleaned in a twinkling and put back again when necessary.

Although, in many places artificial hatching beds may be, and are, most successfully worked without filtering the water, yet there are many streams which bring down such an amount of earthy matter every time there is a spate, that filtration becomes absolutely necessary. When such is found to be the case a very good plan is to dig out a settling pond, or if necessary two or three of them, on the water supply. Do not on any account attempt to make one on the stream itself. Draw the water from the stream and pass it through a hole sunk in the ground, say at least six feet square, and three feet in depth. If twelve feet square and four or five feet deep, so much the better. It does not matter whether the sides are rectangular. Let them be so if practicable, but should the land not allow of it, it is not worth troubling much about. The great thing is to pass the water through a large hole or pool, in which it will have a chance of being fairly tranquil. The result will be that the bulk of the sediment brought down during a spate, will be left at the bottom of this pool, and the water for supplying the hatching bed should be drawn off at or just below the surface. If one pool be found insufficient, have a couple of them ; they are easily made and at a very trifling cost. Where stone is available the sides are better built up of loose stonework, as the action of frost is not so much felt, as is the case where they are simply dug out and left.

To make the thing more perfect still, should it be found

needful, it is well to pass the water through a gravel filtering bed. This is easily made by digging out another hole, say four feet square and three feet deep, and half filling it with coarse clean gravel, free from sand. On the top of this put a layer of fine gravel, about an inch to two inches deep, and on this a thin layer of thoroughly clean sand. The water should enter this simple filtering tank at the top, and be drawn from the bottom of the mass of gravel by means of a pipe. This pipe should be perforated, in order the more readily to take in the water, which, of course, will rise to its own level, and may be drawn off just below that height. Such a filter requires no attention while hatching is going on, after once it is set working, unless the water be very dirty indeed. In such a case, it may be desirable to have two filters, and while the deposit is being taken from one, the other goes on working. Should a filter require cleaning, and be run dry for that purpose, the deposit is found to consist of a thin layer of mud over-lying the sand. This mud may easily and quickly be removed, a little more clean sand added, if needful, and the whole is again in working order. It is, however, hardly needful to have two filter beds, as the water can be so arranged that it can be shut off the filter, and yet kept on through the ova bed, while the filter is being cleaned. The whole cost of making a couple of settling ponds and a filter such as I have described need not exceed twenty shillings. Some fish culturists prefer to reverse the action of the filter, that is, to make the water enter it at the bottom and flow off at or just below the surface. They work very well either way for sufficient length of time to hatch fully eyed ova, but of the two systems I prefer the first. It is much more easily cleaned should necessity require it, and on the whole, is safer in its action than the other.

In some American hatcheries wire grilles are used, and seem to work very well, but, on the whole, the results are not so satisfactory as those obtained on glass. The iron wire of which they are usually made has to be well coated with varnish. My experience of varnish is that from various causes it comes off and leaves the metal exposed This necessitates great care in the handling and use of these grilles, and at one large American hatchery which I visited I found that as soon as the eggs would bear removal they were placed on fresh grilles, whilst the others were being dried and re-varnished, ready for further use. Apart from accidental scratches, which let in the water, it will percolate through and corrode the metal, and wherever this corrosion takes place the eggs suffer. One very weak point I have found to be the place where the wires enter the wooden frame. At this point it is often difficult to prevent corrosion taking place during the period of incubation, and, wherever it does take place, there it causes injury to the delicate embryos. It may not kill them at once, but it weakens them so that they cannot live to grow up.

Livingstone Stone says—" Fourteen trout eggs were placed on a copper-wire screen in November, 1869, at the Cold Spring Trout Ponds, and in fifty days they had absorbed so much copper that they were of a dark brown tinge and hard like peas." Many of my correspondents have found metallic trays, chiefly zinc, very hurtful indeed to the ova.

It is better to work with fewer eggs, and to do the work well, than to go in for large numbers; and I would hand on this piece of advice to all who contemplate having anything to do with the incubation of ova. It is the key to the whole work, and any point overlooked, however trivial it may appear at the time, may cause wreck and ruin afterwards. The preparation and incubation of the ova is a special work, requiring much care and attention, and can only be successfully done by those who thoroughly understand it. The hatching of ova, after having been properly prepared and incubated, is a very simple matter indeed, and can be done by any man of ordinary intelligence. So simple and easy has it been made, indeed, that the eggs, as already explained, will hatch themselves, if placed in a well-made artificial hatching bed. It is here that we gain considerably on Nature. Nearly all the loss which takes place in the egg stage is prevented.

Loss has been variously estimated by different specialists, but we are not far from the mark in assuming that a very small percentage, indeed, of the eggs naturally deposited in our streams live to produce mature fish. The number probably varies a good deal according to circumstances, but we know that in many cases not one egg in a thousand survives. Whole spawning beds are sometimes washed away, and the contents destroyed. The host of creeping things found at the bottom of our streams, almost without exception, prey extensively upon trout ova. Every fish culturist has found these creatures at some time or other coming down into his hatching boxes. The filters will not keep them out. They are so minute, in their early stages, that they are practically invisible, and they get in unobserved, and then grow so fast that soon they cannot get out again. In a well managed hatchery, however, the trouble arising from this cause is trifling. But in a stream, matters are very different. There are creatures innumerable, all instinctively attracted to the place where the eggs are deposited—caddis worms, creepers, shrimps, beetles, frogs, mice, rats, ducks and other fowl, eels, trout, grayling, salmon, sea trout, and nearly every other fish found in the streams. Nearly everything ranks as an enemy to trout ova. Therefore it is apparent, that by protecting the eggs, we are doing a great deal. But we are doing more, for the artificial beds shelter the "alevins" for awhile, and after they drop down the little artificial stream they are still shielded from the bulk of the dangers enumerated, especially from the depredations of trout and other fishes.

The streamlet used as a hatching bed should be so constructed that no outsider of the trout family can by any possibility get into it. Eggs, before hatching, have no power of protecting themselves, or of getting out of the way of danger. As soon as the embryo is out, it at once possesses some power of self-protection. It has a pair of well-developed eyes, and knows how to use them, and immediately that wonderful power called instinct causes it to seek a hiding place. So strong is this desire to hide, that if the little creatures cannot find any other place they will hide under each other, and in doing this they gather together in dense masses, reminding one of a swarm of bees.

Californian baskets, which are simply wire cages in which eggs are piled one above another, do not suit the eggs of our British *Salmonidæ*. It is true that by using them a large number of eggs can be hatched in a small space, but the result is undoubtedly a partial suffocation of the embryos, and a general weakening of the young fish. They may do well in America, but what will suit the fish of one country often may not agree with the fish of another. A lower prevailing temperature is, no doubt, to some extent, more favourable for the use of such an apparatus, than the comparatively higher one in this country. But whatever be the reason, I am satisfied of one thing, and that is, that the grille system is the one of all others for successful work in Britain. It gives the eggs plenty of room, and removes all danger of suffocation where a good current is run over them. The carbonic acid given off by the egg is at once carried off, and the absorption of the oxygen from the water is unhindered.

In Germany, perforated glazed pottery is used extensively for hatching ova upon, and, with the water welling up through the perforations, it does very good work. It has its objections, however, those of cost and liability to breakage being not the least. A broken tube in a grille is easily repaired, or replaced, but not so with a piece of pottery. To repair a broken tube, all that is necessary is to insert a piece of charred wood, cut to the proper size, into one of the broken pieces, so that half of it remains projecting. Then slip the other portion of the tube over this projecting piece, and the tube is again ready for use. Grilles can be made of ordinary window glass cut into strips, and fitted into a cogged or notched wooden frame. The cost of making the frames, seeing that every cog has to be separately charred, is considerable, and it is on this account chiefly that they have fallen into disuse. A few still remain in operation in my own hatcheries, and considering only the hatching of the ova they do quite as well, if not better, than the more modern arrangement of glass tubes or rods. It has been suggested that the rough edges of the glass will cause injury to the ova, but this is by no means the case. I have hatched many millions of ova most successfully on these grilles, and have found them to work very well indeed. The advantages of the more modern invention are that they are a little easier to make, and the glasses are not so apt to fall out, and those who have to use them prefer them to the old style.

One of the chief things to be avoided in a hatchery is the overcrowding of the eggs. It should never be permitted under any circumstances. I have tried a number of experiments with ova, with a view to economising space when necessary, and many years ago, before baskets and cages were thought of, I invented a hatching-box, in which several grilles can be placed one over the other. As soon as the eggs will bear moving, the grilles may be lifted from the hatching-boxes and placed in a compact storage-box, the working of which is explained by the accompanying woodcut. This box is divided into compartments, each holding four grilles, or, if the grilles be placed directly on each other, six of them may occupy the same space. The compartments, it will be seen, are so arranged that the water enters each at the bottom, rises up through the grilles, and keeps the ova in sound healthy condition. This box I designed in the early days of fish culture, and have since used it successfully both for ova and fish. It was honoured by a medal at the Yorkshire Exhibition held at Leeds in 1875, where the design was applied to a series of aquarium tanks, for which purpose it answers admirably. I had it in use in 1870, and about the same time it was patented in America.

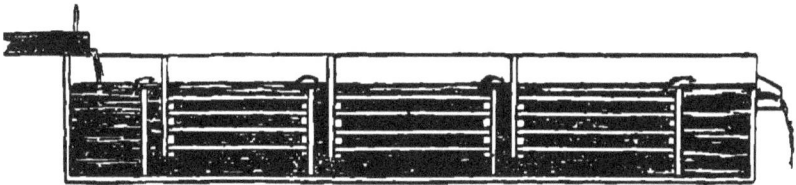

Fig. 20.

CHAPTER IX.

Ova to the Antipodes—The tropics—Various methods—Modus operandi at the Solway Fishery—Selecting and preparing the moss—Its cultivation—Woven fabric —Best time to pack—Ova hatch rapidly on unpacking—Long voyages—Unpacking —Washing off the moss—Fully eyed eggs.

I T is now pretty well known that trout and other ova can be packed and sent with success, not only to any part of the British Isles, but to any portion of the world that may be desired. It has been so sent, and the waters of the Antipodes have been most successfully stocked, as we have seen, by means of ova sent out first from this country. Attempts are being made to stock some of the hill streams of the tropics. Up on the mountains, where the atmosphere is cool and the cold snow water comes down the streams, trout should succeed well, and it is by means of ova that they must be introduced.

The use of ova, too, is destined to play a very important part in the stocking of our waters at home, and therefore the packing of the delicate little morsels becomes a matter of very great importance. I find it is rather a common notion that trout eggs can be packed up by anyone at a moment's notice, and with very little trouble. A greater mistake could not be. The successful packing of ova is an art that has to be learned by careful training and experience. I have, in the course of my life, received a great many consignments of ova from a great many different people, both at home and abroad. When I say that no two senders have packed their eggs exactly in the same way, it will be apparent that there is a great diversity of idea on the subject. I have seen eggs packed in mosses of many descriptions, in cotton wadding, in flannel, in water, in gravel, in muslin, and many other substances. The work is by no means difficult, and when the principle is once understood and the requirements of ova have been sufficiently studied, a dexterous hand will soon become expert at their manipulation in large quantities. When a quarter of a million of eggs have to be packed in a couple of hours a system is required in order to carry on the work. The eggs are packed in trays which are made of wood with a bottom of perforated zinc.

One worker places a sheet of felted moss at the bottom of each tray, and upon it a piece of swansdown or fine netting. Another takes the eggs off the grilles in the hatchery, and carries them into the packing room, where they are deposited in bowls, on a bench near the packers. The eggs are so easily counted on the grilles that the exact number in the bowls is known. They are all measured as they are put into the trays, however, and the measure being ascertained by counting to hold a certain number, checks the count of the ova on the grilles, and prevents the possibility of any mistake.

The eggs are transferred from the measure to the fabric in the packing trays, and by a stroke of the hand gently spread with the help of a heron's feather. The tray is then handed to an expert, who picks out any blind or white egg, or puny embryo, should such be discovered to have been overlooked when the eggs were picked over on the grille in the hatching box. This can only be done in the daylight, as it is impossible to detect the difference when eggs are packed during the dark hours. The layer is then covered with another piece of fabric, over which is placed a second layer of felted moss, which in its turn receives a sheet of fabric, and a layer of eggs. Three layers of ova, four or five sheets of felted moss and six egg cloths form the contents of each tray, and the trays are piled one on the top of each other as required, up to the number of eight, and these are placed in an inside case contained in a box of sawdust. They are usually lifted out of this case by means of a flannel band, which is passed underneath them in packing. In this way a large number of eggs may be packed in a short time by either men or women. They should never be touched by the hand.

There are other ways of manipulating them, and I suppose that at every hatchery the plan will vary somewhat, but, after trying many other methods, I have found the one described to be the best and simplest. Two thousand eggs are placed on a layer very often, and in this way forty thousand eggs occupy the same space as a much smaller number would, thus saving labour and freight to ourselves, and a lot of trouble and half freight on empties to the receivers. I have carefully tried many experiments with ova in order to ascertain the necessary conditions for conveying with safety the largest number of eggs in a given space of reasonable dimensions.

A great deal more depends upon the packing, the temperature, the supply of oxygen, and the preparation of the moss than most people suppose. To begin with, then, let us consider the moss itself. I tried a good many kinds of moss, and there are several kinds that under proper cultivation may readily be made to answer all the purposes required. But to gather mosses indiscriminately, taking any likely-looking stuff that comes to hand first, is not the way to promote the safety of the eggs. Many mosses grown in woods are unsuitable, from the fact of their being so much of foreign admixture among them in the shape of minute bits of stick, rotten leaves, roots (some of which may be highly poisonous), and other matter. Those found in very wet places often grow so luxuriantly that the lower parts decay, and are on that account not good for packing ova. I have found some sphagnum beds to be full of animal life, and others again containing *Saprolegnia*, or fish fungus. In a large establishment, where everything has to be systematized, the moss is felted, and as only some kinds can be readily manufactured into felts, the moss question becomes a very serious one. So much is used at the Solway Fishery that it has become needful to grow it, and this is an excellent way over the difficulty, as it can be gathered clean and fresh, there being only one year's growth to deal with. After it is gathered the ground is replanted, and although no manure is applied, and the same crop is taken off it each season, yet after several years the soil seems to be in no way impoverished. The only change that has been observed is that each succeeding season a finer and richer crop of moss is produced.

It should be well adapted for packing embryo trout, for it has been largely used for packing human infants by the Lapland women, who wrap their children in it. When well dried it is found to provide an exceedingly good protection against cold.

The moss is gathered and carefully kept in a cool, damp place, where it undergoes the felting process, which is very simple. A number of presses are filled with layers of moss and sheets of perforated zinc alternately, and after remaining all night, the moss is taken out beautifully felted next morning, and so a fresh supply is continually being produced. It is not by any means necessary to felt the moss; indeed, so far as it and the eggs are concerned, it answers the purpose just as well without being felted, but for packing large quantities of ova it is an absolute necessity to have everything expeditiously done, considering the price at which trout eggs are put upon the market.

The fabric which is used between the moss and the eggs should first be well washed, to cleanse it from starch or any other impurities. I once received a lot of ova which may be said to have been packed in window curtains and starch. The sender had cut up an old curtain, and used the pieces stiff with starch for packing the ova; with the result that on unpacking here they came out a conglomerated mass of eggs and starch. The eggs were covered with the slimy stuff, and had to be well washed. I have seen more than one case of this kind. Now, although starch is considered sufficiently wholesome to give in large quantities to human infants, it is decidedly bad for trout, and its use is to be avoided. Had the journey been a long one the eggs might have suffered materially, and in any case there would be a great risk of their being attacked by that most dreaded pest of the fish culturist, *Saprolegnia* (fungus). It would not attack them probably until they were in the hatching boxes, as it does not grow readily in air, too much oxygen being fatal to it. Hence the necessity for carefully washing the eggs before laying them down. It should always be done.

Old window curtains, after being well washed, make excellent packing material for trout ova, but a variety of fabrics may be used. The great point to be observed is cleanliness.

Eggs may be packed as soon as the eye spots begin to show

black, and, with the rapid steam communication which we have nowadays, they may be safely started on very long journeys at this stage. They are about half incubated, and, when packed in moss and iced, they will not hatch so soon as they would have done in the trays in the hatchery. The low temperature at which they are kept during ocean voyages slightly retards the development of the embryos. When packing eggs for export a few are invariably retained, packed in moss exactly like those exported, and kept at a low temperature, and these usually remain unhatched for fourteen days or so after the date of arrival of the consignment sent abroad. On being unpacked and placed in a hatching box they generally hatch very soon.

I have received accounts of ova successfully sent to the Colonies, which have, on unpacking and being placed in the hatching boxes, hatched in a few hours. On one occasion a lot which were sent to Natal hatched off in a few minutes, the temperature of the water being over 60°F. This was a very natural result after being taken from their well-iced packing case. The temperature of the water was first lowered by means of ice, and then allowed to rise gradually. Ova can now be sent very successfully to any part of the world. It is not to be supposed that no loss takes place on these consignments. Sometimes they travel beautifully, and, with the well-studied system of packing that is now adopted, there is every chance of this, provided they are properly attended to during the voyage out.

For long voyages the eggs are packed in a very similar manner to the one adopted for home packing, though a few extra precautions are necessary. An air space is left at the bottom of the packing case, and the perforated zinc bottoms of the trays are all double, a quarter of an inch being allowed between the double bottoms for ventilation. The zinc used is never new, but always well seasoned material that has been used for screens between the fry ponds the year before. This does not dissolve and produce any salt of zinc, which is fatal to ova, and should be carefully guarded against. An empty tray is fitted into the box at the top, which is filled with ice, and kept full during the voyage, and should any hitch occur here, especially going through the tropics, all the consignment will be lost.

Then, again, care has to be taken as to the moss that is used. It must be perfectly fresh and in growing condition. It has been noticed that when the moss goes bad or ferments the eggs in that tray are all dead. When it retains its vitality, and in reality grows a little, the eggs travel well, other things of course being equal. The ice in the uppermost tray melts slowly, and the water percolates through the moss in the trays beneath. But for the ventilation given, the delicate embryos would be in great danger of drowning. It was found for some time that the eggs in the bottom tray suffered the most, and were often in bad condition. On this account an air space was left at the bottom, and this also serves as a temporary receptacle for the water caused by the melting of the ice above. This water does not remain here, however, but escapes through a hole in the box bottom, and the box itself stands on a couple of wooden slats, one on each side, which act as feet and allow the access of air, and the same time the escape of the water.

The eggs may be packed between layers of woven fabric or of mosquito netting, but this is in itself a source of danger, as decay sets in, and affects the vitality of the embryos. It may not prevent the eggs hatching, and in some cases may not do them harm, but they remain in a more healthy state during long voyages, when packed between layers of felted moss only. The unpacking is a little more difficult, but the eggs turn out better, and after all, with a little practice in unpacking, there is no difficulty. The two layers of felted moss are kept apart by the eggs which lie between them, and are therefore easily separated. After lifting the top layer of moss a piece of linen fabric may be laid over the eggs. The tray may then be inverted, and the eggs will mostly remain on the cloth when the tray with its contents is righted again. Any remaining eggs may be feathered, or otherwise picked off. Should a layer of felt fall to pieces during unpacking, as may easily happen, it is readily separated from the ova when all are placed in a bowl of water, as the specific gravity varies considerably. The eggs will go to the bottom of the bowl, whilst the particles of moss are held for a time in suspension in the water, and may be poured off with it, and then a little more water added and poured off again, the process being repeated until all the eggs are clean.   Ova packed on layers of linen fabric may on unpacking be similarly treated.   Take up the cloth by the four corners with the eggs upon it.   Immerse in a bowl of water, let one side go, and gently withdraw it, and the eggs will float off. Should a few adhere to the cloth feather them off.

Even on the grilles the moss may be washed off.   I have seen a lot of eggs successfully laid down with much moss among them.   A strong current of water being turned through the hatching boxes, the moss at once floats off, and is caught on the screen near the outlet of the box.   The work is quite easily done with a little practice, but the beginner must not expect everything to go just right at first without any trouble.   I have seen individuals who made great bungling at the work at first soon acquire an amount of dexterity, which enabled them to manipulate large quantities of ova successfully.

Trout ova may be packed and sent anywhere in Britain, or on to the Continent, immediately after impregnation, provided the journey does not exceed forty-eight hours, and they are travelled at a low temperature.   The best time of their existence for such journeys, however, is when the embryos are well developed, and the eggs are in the stage called "fully eyed." They are then near hatching, and in a mild temperature should hatch in a few days after being received.   When very close to the hatching point they may sometimes be sent with safety by using ice, and even later in water in bottles quite successfully.   The latter plan is not to be recommended, but may often be safely adopted in an emergency.

CHAPTER X.

Word derived from the French—Appearance on first hatching—Very helpless at first—Begin to pack—Hides to be avoided—Provide lids for the boxes—Structure of alevins—Cleanliness—Guard against rats or mice—Water insects—How to detect their presence—Cripples—Deformities—Dropsy or blue swelling—Constitutional weakness—Fungus—Paralysis—White spot—Suffocation—Still waters.

FISH culturists have apparently by common consent adopted the French word "alevin" as the name for a newly-hatched fish, so long as it has the yolk bag or umbilical sac attached to it. The length of time varies much in different fishes, in some being only a few days, in others many weeks. The temperature of the water in which they live has a great deal to do with the absorption of this sac. It has been found that in comparatively warm water a trout will absorb it in a little over three weeks, whereas, when it is very cold, sometimes more than three months is occupied in the process.

When the eggs are hatched the little fish or alevins slip through between the glass bars of the grille and go to the bottom of the box, leaving behind them a mass of empty egg-shells. The grilles may then be lifted out, washed, dried, and put away for next season's use. When the shells are all removed, as described in my chapter on "Hatching the Eggs," we may turn our attention to the delicate-looking little beings which the hatching box now contains. At first they lie panting on their sides, and if disturbed they will make a start, apparently in any direction in which their heads are pointing, and entirely regardless of consequences. After a few vigorous movements through the water they suddenly stop in their headlong career and quietly settle to the bottom, where they again lie, panting and apparently out of breath. At this time of their existence they are helpless creatures indeed.

When hatched naturally in a brook or in an artificial ova bed, the case is very different. They are then amongst the gravel, which is the protector which nature gives them from their enemies. The hatching box is perfectly clean, and there is, or should be, nothing which they can get under. They will very soon begin to make up for this by getting under each other, and in their efforts to do so will collect in dense masses which have been likened to swarms of bees. In fish-cultural language they begin to "pack," and there is no more healthy sign than to see all the alevins in a box well "packed."

Many years ago the great idea was to provide some artificial "hide" or place for shelter for them, and many devices were thought of, but all proved to be great failures. The effect of placing a stone or other object in a hatching tank for the fish to get under is to provide a place where there is no current, and which instead of being beneficial will only prove a "death trap" to a large number of the alevins. By leaving them alone, however, a very different result is obtained. Some of them begin to feel an instinctive desire to get into a place where the water is constantly changing, and having found out by following up the current the exact point which best suits them they remain there, and others gather to them, and before long they will all be densely packed with their heads turned the same way, and their noses pointed downwards.

The sight of from ten to fifteen thousand fish in such close proximity to each other might lead one to suppose that many would be suffocated, but this never proves to be the case. They are perfectly safe as long as they remain packed, but should they begin to scatter at too early a stage of their existence, then look out for mischief. Whilst packed they are just in that particular part of the current which suits them best, and as long as they remain there they will take no harm. I am, of course, assuming that the water supply will in no way be tampered with just at this time. A slight disarrangement of the current will cause the little fish to move, or, as one of my men used to put it, to "shift their quarters." And this is better avoided. Too much light is also very disturbing to them, and the box lids should be kept closed except at the ends. The front end of a lid should be placed on the water board or breakwater at the head of the box. This will prevent any light getting in from that direction. My lids are half the length of the boxes, that is two lids to a box. By removing the lower lid, therefore, that half of the box is exposed to the light, and the alevins soon desert it, and they push their way up to the darkness, and it will be seen that the darkest portion of the upper half of a semi-covered box is that part which lies a little behind the water board. Here the alevins are not only out of the light, but they are also in the place of all others where the currents are best adapted to their requirements. By slightly dis-colouring the water at this point, by means of some harmless ingredient, the currents can be quite easily traced.

So much, then, for the packing of the "alevins." Let us take another good look at one of them soon after escaping from the shell in which it has been developed. It is possessed of a body having a head in which are placed two large eyes. The mouth is apparent, but requires further development. There is a large transparent looking sac attached to the fish, which contains a number of yellow oil globules, and is covered with a network of veins, and the heart may be seen, showing as a deep red patch of colour which pulsates regularly. By using a microscope, a more interesting sight may be witnessed than in examining an egg, inasmuch as the little creature is now further developed. The heart is to be found well forward of the umbilical sac, or between it and what will be the lower jaw ; it is double, one side of it receiving the blood from the veins, and the other sending it out again to all parts of the body. Two blood vessels will be seen to run the entire length of the body, and the blood may be seen coursing along within their walls, in the one going towards the heart, and in the other in the opposite direction. One of the main arteries is also to be distinctly traced by the naked eye on the underside of the sac, and the action of the blood may be clearly seen here also. I do not know a much more interesting microscopic sight than that of the circulation of the blood in a young fish at this stage of its existence.

Of the fins the pectorals are the most developed at this

o

stage and are kept in constant motion. They are rather suggestive of a couple of miniature fans, and they have a some-what similar and a very important part to play, in causing fresh currents, and so rendering assistance to the gills. As a lady's fan is used in a close and heated room for producing currents of air, so these little fans or "pectorals" are used for producing currents of water, and they form no insignificant part of the breathing apparatus of the little fish. They not only assist respiration in this way, but also by driving off impurities which may exist, as for instance those caused by exhalation from the fish themselves.

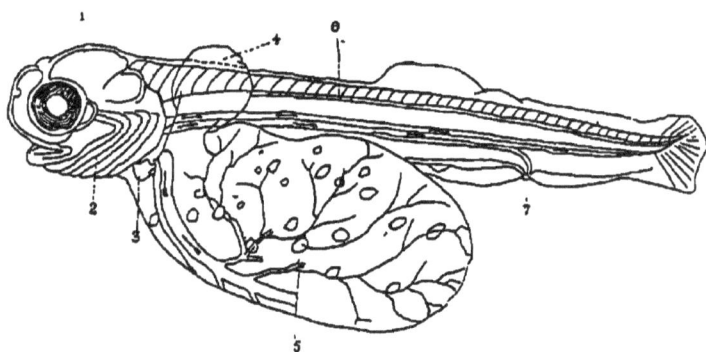

Fig. 21.

The accompanying figure represents an alevin trout magnified. (1) The cranial cavity containing the brain ; (2) the gills or breathing apparatus; (3) the heart; (4) one of the pectoral fins, used at this stage for assisting respiration by causing currents, and acting like a fan to the gills; (5) here the wood-cutter has not connected the dotted line, as he should have done, with one of the numerous rounded bodies seen in the umbilical sac or yolk bag, and which are oil globules: he has also very cleverly made this dotted line to close the chief vein, making it appear as if cut off, whereas, in reality, it simply disappears from view at this point by being out of focus ; (6) represents the vertebral column ; (7) the anus. The rudiments of the dorsal fin are quite apparent, and the rays of the caudal fin or tail may also be traced. The sketch is taken very soon after hatching, and the daily development of the fish from this point is a most interesting study, and well worth the attention of anyone who is interested in such matters.

The hatching boxes should be kept as clean as possible during this period, for although the alevins themselves will scour the floor and keep clean the space immediately around them, yet they are liable to be very seriously incommoded by dirt in the water. Anything of a fibrous nature is apt to be drawn to the fins, in a way which must be exceedingly uncomfortable to the fish, even should it not in the end prove fatal.

During the "alevin" stage of the trout's existence there should be very little trouble as a rule in the hatchery. The daily watchfulness as to the water supply, cleaning of filters, and picking out any dead alevins that may occur, forms the chief work, so far as the young fish are concerned. A sharp look-out should be kept for intruders in the boxes, for it is quite possible that some creature may find an entrance in some unlooked-for way or other. Rats I have already mentioned, and mice are in a degree almost as bad, and they are much more difficult to keep out than rats, as they can get through such a much smaller hole.

In the early days, when my hatcheries were not vermin-proof, a very sharp look-out had to be kept, and recourse had to constant trapping, but, notwithstanding, it was found absolutely needful to make the buildings thoroughly proof against these enemies. After this had been done, as I thought, still they came and it was a puzzle to find out how they entered. At last it was found that they came through what seemed an impossible means of entrance, immediately over one of the doors. This was stopped with broken glass and cement, and for a few days all was clear. Then they came through the roof, but this means of ingress was also stopped, and now the only chance to get in is by the door, when it is open. I say, *when it is open*, because it is carefully kept closed, except at times when anyone is passing in and out. There is an excellent trap which I now use, and which I will guarantee will keep any place clear of rats if properly worked. I will describe it later on.

There is a possibility in some hatcheries of the larvæ of water-insects getting into the boxes, and they are destructive. Their presence is easily discovered, though. Some morning, in looking over the stock, a number of white specks may be noticed in one or other of the boxes, consisting of precipitated albumen. When this is observed, look out for the cause of the mischief. Some one or other of these water-creatures has obtained admission, and the sooner he is ejected the better. He is all right enough in his proper place, but that is not amongst "alevin" trout. He may turn out to be a caddis worm, and if so, will probably be found lurking in a corner. He has most likely come in when much smaller, and then grown bigger. He may well do that where he is, for he is certainly in a water of plenty. In order to ascertain how he goes about his work, just place him in a glass jar containing only water, for twelve or thirteen hours, and then introduce an alevin trout. He will soon make up to it, and with one squeeze of his powerful claws, or nippers, will burst the umbilical sac, and devour the contents. This accounts for little bits of precipitated albumen occurring in the box.

A few alevins will die, but they should be very few, and these will drop out of the "pack" and lie about lower down in the box, where they will linger for awhile and then expire, first changing to a lighter colour, and becoming rigid. It is not at all unnatural that this should be so ; they are probably weakly fish that never would have lived to grow up, and which always occur in small numbers. Then there will be found a collection of fish which will not die at once, but which still appear very helpless, and usually gravitate to the lower end of the box near the screen. On examination it is easy to see that they are cripples, and very curious-looking individuals they are, most of them suffering from curved spines. This weakness amongst young fish is well-known, and assumes a great variety of forms. Some have their backs bent so as to look sickle shaped, whilst others are twisted up very much after the manner of a cork-screw. In some cases the body is bent at almost a right angle, and in others the tail or the head is deformed. These are soon devoured when their lot is cast in a stream, but here they lie, if allowed to do so, until the umbilical sac is absorbed, when they die. There are also some curious examples of a different nature, and these are seen to be possessed of two heads and two bodies with a tail common to both. Others, again, have a head and half a body extra, the latter being attached to their broadsides. Some have one head but two tails, and occasionally two fish have only one umbilical sac between them. A rarer occurrence is to find one possessing three heads, and rarer still three heads and three bodies with only one tail. None of these live to grow up. They usually die soon after the absorption of the "sac," and although many persons have tried to rear them, no one, as far as I know, has yet succeeded in doing so, with the exception of an occasional cripple, whose body has not been very seriously deformed. I have, at the time I write, one of these moderately-deformed fish which I succeeded in rearing, and which has been spawned for four seasons in succession. The malformation does not appear to be hereditary in this case, and the specimen is a brook trout *(Salmo fario).* I had two fish amongst a batch of American trout, a few years ago, which had abnormally large rounded bodies. They happened to be a pair, and the eggs were duly taken from the one and impregnated with milt from the other. The young fish were carefully watched, but no sign of the deformity of the parents was visible amongst them.

A few alevins will probably be noticed at or near the lower end of the hatching box that have distended umbilical sacs, the part affected looking almost colourless and transparent. There is, in fact, a clear watery fluid, which is discharged, or partially so, on a puncture being made in the outer sac. The umbilical vesicle really consists of two sacs, one inside the other. The inner sac grows less as its contents are absorbed, but the outer one has not the same contractile power. It can be punctured, however, without apparently causing any pain to the fish, but the inner one is evidently very delicate and extremely sensitive. The proper function of the outer sac seems to be the protection of the inner one, and on the final absorption of the vesicle a part of the outer sac often drops off.

The disease in question is generally known as the "dropsy" or "blue swelling," and was, I believe, first so named in America by Seth Green. In many cases, the part affected has a slight bluish or bluish-grey tinge, from which it gets its name. There is considerable reason for assuming that it is often caused by using immature breeders, and it also occurs occasionally in excessive numbers in cases of hybridization. Otherwise, the affected fish are not numerous, and as there seems to be nothing infectious about it, it is not to be much dreaded.   I have tried many experiments with it, and out of a large number of fish whose sacs were carefully punctured only one recovered.   Seth Green was also successful in performing the same operation, but I question whether anything is gained by it beyond the knowledge that it can be done, as such fish will probably never live to grow up, owing to inherited weakness of constitution.

A few others among the alevins will also die from constitutional weakness, although they may not have developed the "blue swelling."   They are, however, like any other weakly animals, ready to develop anything that surrounding circumstances may favour, and amongst others will often exhibit inflammatory symptoms.   Sometimes they are attacked by fungus *(saprolegnia)*. Every precaution ought to be taken to guard against it, and this being done, fungus ought to be unknown amongst alevin trout. The great safeguard to be employed is the same as for the eggs; have all woodwork that comes in contact with the water well charred.   One of the worst features of fungus amongst young trout is that it frequently attacks the best fish.   Next to uncharred wood, nothing is so likely to bring it on as dirty hatching or rearing boxes.   These should always be kept perfectly clean.   I have already pointed out the necessity for this during the period of incubation, and repeat the warning now as it is of the utmost importance.

A very sharp look-out should be kept, and should fungus make its appearance from any cause, an application of sea water will do good.   Where this is not easily obtainable, a dose of chloride of sodium or common salt will prove beneficial.   Or a piece of rock salt may be placed in the water at the intake.   It will have the effect of making it slightly saline, and this kills the germs of the fungus and also the plant itself.   The amount required depends so much upon surrounding circumstances that it is difficult to lay down any fixed rules as to quantities, but for a hatching box twelve feet long by nine inches wide, five or six pounds of common salt is not at all too much.   It may be introduced by simply placing it in the inflowing current.   Of course

it is desirable for anyone performing an experiment of this kind
for the first time to exercise a little care. Do it gently, and watch
the effect it has on the fish. Where the water is right, and all in
order, it will do no harm, but will be highly beneficial. It also
has a tendency to destroy any animal parasites that may possibly
be attacking the fish, even at this early stage.

Paralysis is a disease to which young fish are iable, and its
presence may be known by seeing some of them come out of
the "pack," and lie very still as if dead, at the same time looking
very pale. They may be mistaken for dead ones, but on taking
them up in a dipping tube they will be seen to breathe, and an
occasional spasmodic movement of the tail may be apparent when
disturbed. They may remain in this condition for some time,
perhaps days even, but they will probably die, and if the cause be
not removed great loss may ensue. To prevent this disease
making its appearance, a good current and plenty of oxygen seem
to be the best specifics.

The depth of the water in the hatching boxes at various
periods of development is a point that has been much discussed.
It is difficult to lay down any special rules for its regulation, as
what does not answer in one case has been found to give very
good results in another. Much depends upon surrounding circumstances, such as the nature and quality of the water, and its
temperature. A sufficient knowledge of these matters can only
be gained by experience, but it may be laid down as a general
rule that the deeper the water the greater supply of it that is
required. While a depth of four inches will give satisfactory re-
results with a flow of thirty thousand gallons a day through a double
supply hatching box, as at the Solway Fishery, at this stage the
depth of water must be decreased where the supply runs short. A
fine jet of air injected into the water has proved to be of service
in some hatcheries abroad, but the better plan is to provide a good
supply of suitable water. Where the upper ends of the hatching
boxes can be raised a little, the advantage of a varying depth is
obtained, but this is not always practicable.

There is another disease which sometimes attacks young
trout, and it appears in the shape of a small white speck within
the umbilical sac. It is usually, if not always fatal, but does not

seem to incommode the fish at first. Sooner or later the alevin loses its energy, however, drops out of the pack, and, after lying about for a few days it dies. I have not ascertained the exact cause of this ailment, and, as far as I can make out, it has not been properly diagnosed by anyone. It sometimes attacks a number of fish, and sometimes only a few, and is probably caused by coagulation or precipitation of albumen. Sudden changes and extremes of temperature seem to produce it in the alevins in some way, and it appears to attack what are otherwise healthy fish.

Care must be taken to guard against deaths from suffocation. When in the egg stage, there is, under certain circumstances, a danger of the suffocation of the embryo, so there is a danger of suffocation of the alevin. Such an occurrence should never happen, but as it has frequently happened in the past, so it will happen again unless guarded against. In the old days the loss from suffocation was very great, when a variety of apparatus was used that was most unsuitable for the well-being of young trout. Practically, I never used anything but plain wooden boxes, but I have seen apparatus that suffocated tho alevins by thousands. Tanks of wood or metal, lined with slate or glass, were of common occurrence, and the joints not being tight, the fish would get into the crevices and behind the slates in large numbers. The cause of suffocation in such places is that there is no current. The same applies in any hatching boxes when, for instance, the water is accidentally stopped, or if the supply be too small, although it is true the fish will often live a long time under such circumstances. It is really simply a question of oxygen. As soon as the supply available is done death takes place. A few alevins may be kept in a basin or a tumbler of water for a long time without changing the water, due regard being paid to temperature. I have seen a single alevin placed in a basin live three weeks without a change of water. Several of my friends who have been wishful to try the experiment have hatched trout ova (when well eyed) in a small aquarium, and have kept a few alevins there till the sac was absorbed, and one correspondent was successful in keeping the fry after feeding for several weeks, the only precaution taken being to ladle the water a little twice a day. As a parallel case I may mention having to-day, as I write, heard from an eye-witness of a lot of yearling trout thriving in a railway ballast pond with no stream through it. They were put in in 1893, and in June, 1894, were seen rising freely. Such cases are exceptional, however, but they go to show that where the water is in a healthy state trout may live. Put that water in motion by making a flowing stream of it, and trout would not only live but thrive in it. There may be plenty of oxygen in a hatching box to keep alive a few alevins for some time without a current, but as the number is increased the water supply must also be increased, and where the box has its full complement of fish it must have a full supply of water, or suffocation will soon be the result.

Livingstone Stone advocates "shallow water with a good ripple." I have found this do very well, and especially for American trout *(Salmo fontinalis)*, but I like a moderate depth and a good current for trout *(Salmo fario)* and Loch Levens *(Salmo Levenensis)*. To ensure success with the alevin trout :—

Have a good supply of water.

Take care the boxes are kept clean.

Allow no hiding-places to exist.

Look carefully to the screens, and beware of crevices.

Sometimes the alevins may give trouble by getting on to the screens, and getting their sacs through, which means destruction. The loss from this cause was formerly often great, but now it is comparatively trifling. A few will get on to the screens, but they are mostly weak fishes, that are better out of the boxes altogether. The desire to avoid light may be made useful here, and any that get about the screens may be syphoned out, and only the best put back again. Take care to put them in at the head of the box. To drive fish off a screen or from any part of the box where they are not wanted pour water upon them. They do not like this; and will vacate the situation at once. To make them stay where they ought to be, keep the place as dark as practicable.

CHAPTER XI.

POND LIFE.

*Water full of life—Care required in dealing with it—The rotifera   Rul
for cultivation—Nature's provision for young fish—Daphnia pulex   Cyclops
quadricornis—Cypris tristriata —Arachnida—Notonecta—Corixa—Gammarus
Dytiscus—Caddis worms—Ephemera—Shellfish—Parasites—Saprolegnia.*

WHEN any important fish-cultural question requires to be answered, the safest and surest way of getting sound information on the subject is to send a thoroughly competent expert to make a careful examination into all the details of the case.   This is the way to arrive at some really definite conclusion, and obtain a correct verdict.   There are so many who are ready to denounce as hopeless anything that they themselves cannot understand or do, that if we listen to their cries nothing would ever be accomplished.   Nearly every section of fish culture has its opponents, and but for the persevering efforts of a few individuals nothing would yet have been done towards cultivating the waters of the British Isles.

The important subject to which I am in this chapter about briefly to allude, is one that is now engaging the attention of men of science, who have had their attention drawn to the great importance of its bearing on the welfare of our fisheries.   At certain times of the year some of our waters are found to contain an enormous mass of living organisms, and we will now consider a few of these beings which inhabit our ponds and lakes, or their margins, with a view to the utilization or otherwise of the supply of fish food which Nature has already provided.

The subject is an exceedingly wide one, and it is necessary at the outset to use caution in the course of our investigations, for we have not got very far before we find that we have enemies as well as friends to encounter, and by cultivating haphazard all the aquatic creatures that we come across, we may be doing harm instead of good. There are probably few of the minute organisms which we meet with which are not devoured with avidity by large trout, but there are many which, if introduced into a fry pond, would soon reduce the stock of fish contained therein considerably.

This is one cause by which trout sometimes seem to die unaccountably in the fry stage. I have seen, years ago, thousands of little fish dropping down dead from no apparent reason. It has since been ascertained that these had probably been attacked by parasites.

There are myriads of small creatures, which are highly useful as food for fry, known as *infusoria*. Some of the best known among these are probably the *Rotifera* (commonly known as rotifers or wheel-insects), but they are very minute, and as far as our present knowledge has attained cannot be cultivated on a commercial scale.

Fig. 22
Rotifer, enl'g.

The secret of the successful cultivation of most, and probably all the minute forms of life, lies largely in the protection of them from their natural enemies. The same lines have to be followed as in the case of large game. We must provide for them safe undisturbed breeding grounds, and suitable food. We often find that Nature works out these deeply interesting matters with unerring exactitude, making due provision for everything. Nothing is overlooked. Now, it does not suit Nature's purpose to have great crowds of any individual creature in one place, except in a few exceptional cases, and hence there are many causes which tend to keep in check the superabundance of life, which would otherwise be found existing. The same applies to the fish. We do not find a pool in the natural stream, as a rule, abounding in fine large trout. We know that this is the exception ; but we also know that it can be brought about by artificial means.

So, then, with the more minute forms of life. If we apply the same rules that attach to the cultivation of the fish, we obtain the same result, viz., an enormous multiplication of individuals. Whilst multiplying the fish on the one hand, we must multiply the creatures on which the fish feed on the other, and so the two things work together. Nature has provided them, and we have only to go in and occupy the position that is open to us. A fact that has been observed over and over again is, that where Nature places the fish there are the necessary food supplies, but only up to a certain point. If we wish to go beyond that point we must increase the food as well as the fish.

I have recently been paying special attention to the cultivation of perch *(Perca fluviatilis)*, and on investigating the natural breeding grounds I found that where the ova were about to hatch there were myriads of living creatures, all ready for the little fish to live upon when the time arrived for them to require food. Yet we hear of individuals condemning perch culture, because when the eggs are hatched in pure barren water the little fish all die. There is no wonder that they do. The cultivation of the *Corregonidæ* has been condemned on the same ground. The little fish are very small and delicate at first, and as soon as they lose their sacs they die from the same cause. But is that any reason why we should abandon the thoughts of their cultivation ? All the more reason for persevering in it until the difficulty is overcome.

Having examined some of the waters of the great American lakes, in which the *Corregonidæ* (white fish) are found in very large numbers, I found the lower forms of life exceedingly abundant. It is on these lower forms that the white fish feed, and the very first discovery of this important fact is probably due to Dr. P. R. Hoy, of Racine. He also found that certain forms were parasitic on the white fish, and it is a noteworthy fact that he failed to find any of these in the stomachs of the fish. This certainly does not apply, however, to all parasitic organisms.

There is one very remarkable occurrence, and that is that the excreta of fish produces a suitable *nidus* for the growth of some of these minute forms, and, therefore, what is looked upon as a nuisance to be got rid of may be made a means of promoting fish life. True it is that everything has its use in Nature. We find that decaying vegetable matter provides some of the conditions which foster the growth of these lower invertebrate forms, and we have had some remarkable instances of the effect of this on newly-made artificial lakes. One thing it is necessary to note, however, and that is, that in some cases the organisms partake so much of the nature of their food as to become highly poisonous to the fish, when it, or the water in which they live, is of an unwholesome character. There is here a very wide field for research, for which we want some workers.

As already stated, the plan to adopt for the cultivation of these creatures is, first of all, to protect them from their enemies. As trout are their enemies by nature, therefore they must be protected from them; and this brings us to the point of having separate or accessory ponds in which to cultivate a sufficient amount of food. At Gremaz, in France, the food is cultivated, and the fish are herded from pond to pond, very much as sheep are penned off on turnips, and are moved from one pen to another as the crop is devoured. What I conceive to be a better way, and one that I have adopted for many years, is to grow the food in separate ponds, to which the fish have no access.

Either plan may be equally good, and probably is so. It just resolves itself into a case of which is most workable. In our cold northern climate we must adapt ourselves to circumstances. With these ends in view, an excellent plan, then, is to provide special ponds, well stocked with plants, for the production of the natural food of the trout. In this way a natural supply may be obtained, and maintained, but if once the trout gain access to the food pond, the result will probably be the entire depletion of the stock of fish food.

One of the best animals to cultivate is the water flea (*Daphnia.*) Though so called, yet it is not in reality a flea, although

Fig 23.
Daphnia pulex, enlarged.

in general appearance it bears some resemblance to *Pulex irritans*, its "longshore" namesake. There are some ten varieties, varying in size from three sixteenths to one sixteenth of an inch in diameter. The commonest species is *Daphnia pulex*, which varies in colour and size considerably, according to the nature of its surroundings, and also to the time of year. It is semi-transparent, and usually of a reddish tinge, and swims with a jerky movement; its food consists of small *infusoria* and vegetable matter. It thrives best in moderately still water, and under favourable conditions its rate of increase is considerable, the females usually producing three broods per month. The females are often very abundant, but the males are usually somewhat scarce. Unfortunately frost seems to be speedily fatal, the next year's insect developing from the epphipial, or winter egg as it is commonly called, contained in the parent at death. At Gremaz, in France, this difficulty is said to have been overcome. Each species or variety of *Daphnia* possesses five pairs of legs, whilst the body is made up of eight segments, and the stomach can be distinctly seen.

Scarcely second in importance is another minute crustacean, *Cyclops quadricornis.* In general outline it somewhat suggests a small lobster, excepting the claws, which are substituted by four long horns, from which it gets its name, *quadricornis.* They are found naturally inhabiting very much the same places as the water fleas, but they are very different looking creatures, being of a whitish appearance, and the females may readily be distinguished by two curious appendages attached to the sides of their bodies towards the tail. These are the ovaries or egg sacs, and they are very prominent objects. The males are without them, but the females are by far the most numerous, and both sexes progess through the water by means of a quick succession of jerks. The sexes may be distinguished by the form of the antennæ, which vary in each, as well as by the presence or absence of the egg sacs.

Fig. 24.
Cyclops quadricornis, enlarged.

The great importance of these creatures will be understood, when we consider that it has been estimated that a single female may be the origin of over four hundred millions of its species in one year ; nay, according to a calculation by Jurine, a single *Cyclops* is capable of producing over four billions in the course of a single year. The calculation is based on the assumption that all live, and go on producing, but in reality, such an occurrence never takes place, as there are so many predacious animals which prey upon these crustaceans, that few of them practically survive. These figures, however, have their lesson ; they teach us of the enormous possibilities that exist, with regard to the increase of these minute creatures, where by so called artificial means they can be protected from their enemies, and allowed to multiply enormously. The food of *Cyclops* is produced by decaying vegetable matter, and minute spores. On occasion it also eats *infusoria*, or becomes a cannibal when a favourable opportunity occurs. It possesses but one eye, and projects itself through the water by means of its oar-like feet, which are ten in number.

The third and last crustaceans of the sub-class *Entomostraca* which it is necessary to mention are the *Cypridæ*. They are interesting little creatures, and are abundant in some waters.

Cypris tristriata is the commonest, although there are said to be fifteen recorded species in this country. They are small free swimming crustaceans, and are enclosed in a bivalve shell-like carapace. The mode of progression is by opening their valves slightly, and putting out a series of hair-like processes or cilia, which they move rapidly and constantly, but on the slightest alarm withdraw them and sink to the bottom. In a closely allied genus called *Candona* we find the swimming apparatus absent, and the species, some five in number, are not found swimming about, but crawling on the bottom, or on the plants growing in the water they inhabit. The food of the *Cypris* consists of both animal and vegetable matter, and they are excellent scavengers. I have seen ponds which they inhabit dry up in summer, but the little creatures, nothing daunted, bury themselves in the mud, and remain there until rain falls, and the pool fills up again. Should the mud be thoroughly dried by the sun they perish, but their eggs retain life even under such trying circumstances, and will hatch and produce a fresh crop in a few days, when placed in the water.

Fig. 25.
Cypris tristriata, enlarged.

As we work out the life histories and developments of some of these interesting little creatures, we are forcibly reminded of the language of the Psalmist, " O Lord, how manifold are Thy works ! in wisdom hast Thou made them all : the earth is full of

Thy riches." That creatures hitherto looked upon as insignificant, should be made to play so important a part in the economy of Nature, is at first somewhat startling. Nevertheless the fact remains, and the more we delve into Nature's hidden mysteries, the more closely are we led up to Nature's God.

The provision that has been made for the artificial feeding of fishes, when man requires to use it, is truly wonderful, only waiting development. From the way in which trout devour these crustaceans, one would suppose that they like them better than we do oysters, or they would not eat them shells and all, as they do at every opportunity.

That they (the crustaceans) are an important item to trout fry there can be no doubt, and they seem to thrive under more varied conditions than some other species. The finest are found in water of the highest natural temperature, other things being equal, and some of the largest specimens were observed in a pool the water of which registered 60° F. It was tolerably shallow at the margin but shelving to deep water, and containing abundance of water plants, and *infusoria* of many varieties.

We now come to a very large family, the *Arachnida*, or spiders, some of which live in the water. They are not so valuable from a fish culturist's point of view, perhaps, as are some of the crustaceans, and some of the foreign species will even kill fish and suck their juices. The smaller kinds known as water mites (Fig. 26) are rather numerous. They are usually in full activity of life in April, and they are just the right size one would suppose for feeding trout fry, but as far as we have observed, the fry seem to reject them utterly. There are brown, red, grey, black, and spotted ones, but somehow fry, even when they get to minnow size, do not seem to care much about them. They vary in size, some of them being about as big as dust shot, whilst others are the size of No. 1 shot. They swim with a beautifully even movement, their legs whilst swimming resembling somewhat a wheel, or screw propeller.

Fig. 26.
Water Mites, enlarged.

The red ones *(Arachnida histrionica)* look particularly bright and attractive, and trout fry will often turn and follow them, but, for some unexplained reason, they do not seem to eat them. It is well known that some spiders are poisonous, and this may have something to do with the matter. Livingstone Stone says :—" If a trout not over two-and-a-half inches long strikes at a black spider in the water, the spider will strike back at him, and if he takes a good aim he will kill the trout instantaneously. The little fellow will not go twelve inches before he turns over on his back, and drops down dead." This is what happens in America.

It is a pleasure to turn from the mites to the water boatmen. Entomologists speak of the large ones as *Notonecta glauca*. They are about three-quarters of an inch long, by about hardly a quarter wide, and have two long ciliated oars to swim with, from which they derive their name. The largest yearling trout devour them greedily. The smaller ones fight for .them, but the fry stand a poor chance—they being the prey of the beetles, who mercilessly harpoon them. Fortunately, although common insects, they do not usually abound in great number; in trout waters. They delight in the stagnant, or semi-stagnant waters of ponds and ditches, in both the larval and the mature stages. Here, again, is a wonderful provision of nature for the production of a large food supply for the trout, when the latter have grown somewhat. The insects may be cultivated in special ponds, and swept out with a suitable net as required.

Another genus, the *Corixidæ*, containing some thirty species, must have our attention. Of these *Corixa vulgaris* is perhaps the commonest. In general form it very much resembles a *Notonecta*, but it swims the other way up; that is, with the oars underneath, and it is only about a quarter of an inch long. Some members of the genus attain a length of half an inch. They are very pugnacious, and are carnivorous in their habits, eating decaying animal matter, and attacking any living creature that is not strong enough to resist them. They swim by means of rapid strokes of their oars, which gives them a jerky movement. They are air breathers, coming to the top, of the water for an instant every few minutes. When the fry are two months old, or even before if vigorous, they hunt *corixa* like a pack of hounds, and eat them

P

piecemeal, as they are quite helpless when once an oar is disabled.
Older fry hunt them alone, and seem to esteem them a great
delicacy.　They take short flights over the water in bright weather,
from April to July, during the breeding season, and will leave
their quarters during a drought, and fly to other waters at any
time that they may be compelled to do so.

They are very prolific, laying a number of small round whitish
eggs, which in size and appearance somewhat resemble those of
the mackerel.　They may be found in numbers on the undersides
of the floating leaves of aquatic plants, and also on pieces of wood,
and on dead leaves in the water.　On the whole it is an insect
that may be looked upon with some favour, although it may kill
off a few of the more weakly, or it may be sickly fry, when it gets
a chance.　It is useful as a scavenger, and as soon as the fry have
grown a little they will not leave a *corixa* alive within their
borders.

When the fry have advanced so far, that they can thoroughly
master a *corixa*, they turn their attention to, and are large enough
to take, a fresh water shrimp *(Gammarus pulex)*, which is
a crustacean of the sub-class *Edriophalina*.
This is one of our very best friends, which should
be introduced by all means into all trout waters,
and cultivated assiduously.　It may easily
be distinguished by anyone, and in general
appearance closely resembles the sand hoppers,
so common on the sea-shore.　It is a very
prolific breeder, and retains its eggs until they

Fig 27
(Gammarus pulex)

are hatched.　It is an excellent scavenger, and appears to be
harmless, although slander says it likes fish eggs.　It does not
appear to do any harm to trout ova, for the experiment has now
been repeatedly tried, by placing *gammari* in the same tank with
trout eggs.　Although their movements have been carefully
watched, they have not been detected doing any harm.　They
may hide under them if they get the chance, but, seeing that in
natural waters trout eggs are buried in the gravel, such an
occurrence is not likely to take place.　In the hatching boxes, of
course, they can do no harm, even if so disposed, for the simple
reason that they are not allowed to enter.　It is their nature to

hide, and this they do under stones, or water plants, or anything
that comes handy, and they feed chiefly at night.

The most favourable places for them to thrive and increase
seem to be shady, overgrown, shallow streams, where they are
seldom disturbed, and in such places they multiply very rapidly
indeed.   I have met with them, too, in reservoirs and lakes,
where they also seem to thrive ; and in open brooks they usually
find plenty of shelter amongst submerged marginal plants or their
roots, or under stones, etc.   Aquatic birds generally know where
to find them, and should not be allowed to enter any places where
gammari are cultivated.   In open streams and lakes it does not
matter so much, as they can there take care of themselves.   It is
most suggestive of shrimps to see the way in which a water hen,
for instance, goes poking about in their hiding places among the
marginal grasses, or the stones at the head of a shallow.   Decaying
animal and vegetable matter constitutes the food of this crustacean,
and it seems to thrive equally well in cold spring water or in that
of a higher temperature, though it usually attains a rather larger
size in water that warms up well in the summer.

Fig 28. *Dytiscus marginalis* and larva

There are a great many aquatic beetles, but they do not
possess a very high food value.   They are hard cased and exceedingly voracious, and probably destroy far more food than they
yield, so there is no economy in encouraging their presence.   The
two largest are *Dytiscus marginalis* and *Hydrophilus piceus*.   The
former is carnivorous and very fierce, and will attack and kill fish
up to the size of minnows, both in its larval and perfect form.   As
a beetle it is well known, being a conspicuous object, but in its

larval stage it is not so often seen, but will be readily identified by the accompanying illustration. As will be seen, it is segmented, and when fully-grown may be one and a half inches long, and is provided with a most cruel pair of jaws, resembling pincers in general form. It is not a particularly quick swimmer, but hunts from ambush, amongst the sub-aquatic plants, and is one of the most destructive insects we have in our fish ponds.

The perfect insect or beetle is an inch or more in length, and appears brown or purplish according to the lights, with a conspicuous yellow marginal line. The under side is reddish brown. The only safe place ·for an individual of this species is among mature trout, which quickly and unceremoniously attend its funeral. As it leaves the water at will, and flies about in the air, and is nocturnal in its habits, we never know exactly where we have it, and it may come any night into our fry ponds. When, therefore, part of the torn body of a little trout is discovered some morning, it is time to keep a sharp look out for the enemy. These creatures sometimes account for a good many fish. We hear of cases in which trout fry in nursery ponds lessen in numbers unaccountably. The *Dytiscus* may be, and often is, the culprit.

Fig. 29. *Hydrophilus piceus* and larva.

Another insect belonging to the order *Coleoptera*, and to the ordinary observer something like *Dytiscus*, is the *Hydrophilus*. It is rather longer, and is more pointed "fore and aft," and is harmless, being a vegetarian in diet, but it is not prolific, and is, therefore, of no great apparent importance to the fish culturist, so far as we know at present.

There is another beetle which must be mentioned in passing,

as it is one that is well known to all observers, and that is the little whirlygig beetle (*Gyrinus natator*), so called because it swims round and whirls about on the surface of the water. It is rather curious in structure, an excellent swimmer, and provided with top and bottom eyes, so to speak, so that it can see either a bird above or a fish below, if they approach with hostile intentions. It also emits an oily fluid if attacked, and has a very disagreeable smell. Its food, so far as we have observed, appears to be flies and gnats just emerging from their larvæ on the surface of the water; at all events it pounces on these with avidity.

We may now consider a few of the flies that spend their larval period in water. They form in most cases a very considerable portion of the natural supply of fish food. Two of the chief families of such flies are the caddis flies, or *Trichoptera*, and the *Neuroptera*, or nerve-winged flies. There are some two hundred varieties of caddis flies, or "water moths," as they are sometimes called, one of the largest and most typical being the *Phryganea*

Fig 30. Caddis Fly, *Phryganea grandis.*

grandis of the entomologists, or the "brown owl" of the angler—so called because of its brown russet colour and soft downy wings. The caddis flies lay a considerable number of eggs, sometimes in the water on aquatic plants, and at other times on marginal plants, whence the young larvæ, on hatching, have to find their way into the water as best they may, many doubtless perishing *en route*. Once in the water the larvæ clothe themselves with bits of stick, stone, rush, leaf, sand, shells, or anything else they come across, forming a tube which serves the double purpose of a protection from their enemies, and also an ambush in which they can approach their prey unobserved. Their food consists chiefly of animal matter, dead or alive; they are capital scavengers, and are excellent trout food, but the larger varieties are great enemies to the young fish, and they are provided with a pair of most formidable claws, with which, if they once get hold, they seldom relinquish their prey until they are satiated.

By way of experiment, six alevin trout were placed in a small tank along with two caddis worms, six with two large beetles, and six with two sticklebacks. The experiment was repeated several times with the same result; the caddis worms killed more alevins in a given time than either of the others. Again, a tank was stocked with a mixed collection of insects, larvæ, small fish, etc., and the conclusion arrived at was, that the large caddis worms are the most destructive of all. The large beetle larvæ (*Diti cus*) is a good second. They each settled minnows and sticklebacks, by gripping them from below. The caddis worms seem to be more active and more persistent in the chase, and their cases afford inconspicuous covers, constructed as they are from bits of their natural surroundings. The moral of this is that it is unwise to have any but very small specimens in fry ponds.

The May flies, or *Ephemera*, with which anglers are so familiar, and which seem in some localities to be steadily decreasing in numbers almost annually, spend three years in the larval form in water. The perfect insects may be readily distinguished by their long tails and vertically folding gauzy wings. The larvæ live in burrows in the muddy parts of the river bank, under water, and it is quite possible that this habit of burrowing may account for their extermination, by the increasing pollution of some streams. The very lengthened sojourn of these insects in the larval stage greatly tends to reduce

Fig. 31 Green Drake, *Ephemera vulgaris*

their importance to fish culturists. This is no reason, however, for not planting them in districts where they have not hitherto occurred, and to which it may be deemed expedient that they should be introduced.

The stone flies, the alder fly, the yellow Sally, the Spanish needle, the willow fly, and the gauze wing, are all well-known friends of the angler. They lay enormous quantities of eggs, which may be found on some of the marginal plants, grasses, etc. The larval stage of these insects is passed in the water, where they are free swimming crawling creatures, predatory in their· habits, and do not possess a protective shell of any kind. Most of them thrive best in running water, and if once introduced, and the water suits them, they multiply rapidly. The various methods of introducing flies to unstocked waters are well worthy of consideration. The work, like most others, has its difficulties, but I apprehend that there is no reason why these apparent difficulties should not be bridged over. Many of the larvæ are very sensitive to changes of water, and this is one of the difficulties that we have to contend with. It can be overcome by using care in transplanting. The creeper, or larvæ of the stone fly, is very tenacious of life, but even it requires care in its treatment, or it will not always be found to succeed on being introduced to "pastures new."

The alder fly (*Sialis luteria*), the black fly which folds its wings along its back, and spends much time sitting on a fence in the sun, or bobbing about on the water in swarms, is best moved in its egg stage. It lays a hundred eggs or more, which are deposited in neat-looking rows on herbage near the water. Careful observation would soon discover these, and it would be much safer and easier to collect and transmit them than to catch and carry the larvæ, and there would be a much better chance of a good result. The introduction of the perfect insect is a mistake, and is probably useless, unless ample protection is given to them and also to their eggs. In the case of some species it may be done by building an insect house, and having had considerable experience as an entomologist, and bred and reared thousands of insects in this way, I can speak with tolerable certainty as to the result. But there are few persons probably who could, or would, devote the time required to look after such an undertaking, and without proper attention it would be a failure.

Previously to the deposition of the eggs, it may be noticed that both caddis and stone flies carry them in a bunch at their posterior extremities; and what angler does not know the grannom fly, or green tail (?), a brown-looking fellow, who often carries a bunch of greenish-looking eggs behind him. They are exceed ingly prolific, and there should be no real difficulty in introducing them.

One other class of insects claim our attention, namely, the dragon flies, of which there are some fifty varieties. They belong to the *Neuroptera*, and on the whole are a bad lot on a fish farm. Most of the larvæ are very predacious, and the curious mask with which they are provided has in close connection a pair of powerful pincers, which they shoot out to catch their prey. They have also equally powerful jaws, with which to dispose of any insects or little fish they may catch. This sort of thing may be interesting, to some observers, to watch in an aquarium, but anything but profitable in trout fry nurseries.

We now come to a class of creatures that play a most important part in the economy of fish ponds, and these are the *Mollusca*, including the *Gasteropoda*, or snails, and the *Conchifera*, or bivalves. As in other classes in this also, the common varieties are the most important to fish culturists. The commonest water

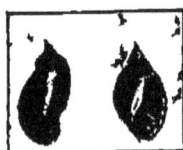

Fig 32 *Limnæus pereger*   Fig 33 *L. auricularius.*

snail is *Limnæus pereger*, and if it should not be naturally present in any water under cultivation, it should be introduced by all means. Even in waters where it occurs naturally, but not plentifully, it is a great help to introduce a few to augment the stock; it is an ascertained fact that with snails, as with larger stock, the breed occasionally seems to get "stale," and almost dies out, and the introduction of a number from elsewhere is a distinct advantage.

For very deep still waters the ear snail (*Limnæus auricularis*) is an excellent one. It grows larger than *L. pereger*, and is prolific.

The trumpet snail (*L. Stagnalis*) is a very handsome and useful mollusc in large waters or ponds, but does not thrive in rivers unless they are deep and slow flowing. It has a long pointed spire to the shell, and is one of the largest of the genus, attaining under favourable conditions, a length of one-and-three-quarter inches. There are a number of others of this class that are both beautiful and interesting to collectors, but for practical importance the first mentioned (*L. pereger*) is decidedly the most

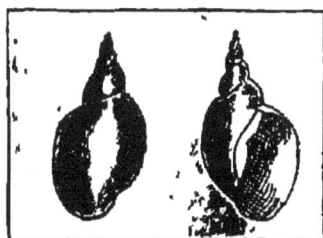

Fig 34   *Limnæus stagnalis*

Fig 35   *Planorbis corneus*

useful. The round flat shell fish (*Planorbis*) are useful and in some waters abundant, but they do not grow so fast nor breed so plentifully as the *Limnea*. There is a useful little limpet, that when once established multiplies rapidly in shallow cold water, where the other snails do not thrive so well. This is *Ancylus fluviatilis*. Fish have often been taken in brooks with these molluscs in their stomachs.

Fig 36.   *Ancylus fluviatilis*

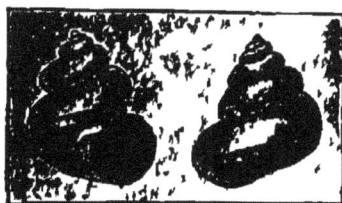

Fig 37   *Paludina vivipara*

The *Paludina* are larger snails, but are ovoviviparous; that is, they keep their eggs within their shells until they are hatched. They are also possessed of a hard horny plate to close the entrance of the shell, called an *operculum*, and the shell itself is thick and strong, and fish will probably not eat them, at least when mature.

They are often found in canals and similar habitats; they are abundant in the Midlands, but several attempts that have been made to introduce them into more northern waters have proved abortive.

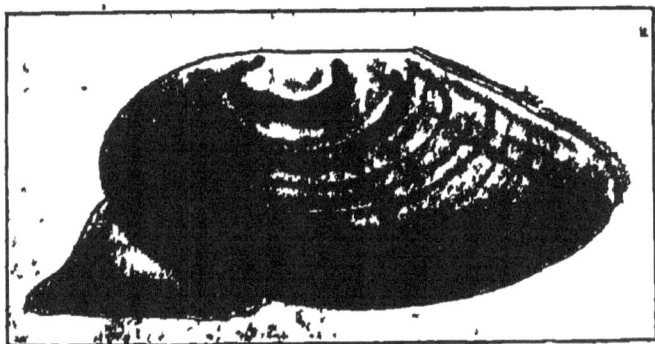

Fig 38 *Anadonta cygnea*

On the bottoms of ponds and streams we should certainly cultivate the bivalves, or *Conchifera*. The largest are the swan mussel (*Anadonta cygnea*) and its varieties. They attain a length of about six inches, and are not of a high food value when mature, but when young they are eaten with avidity by many fishes. In their infancy they are parasitic on fish, but their attacks do not

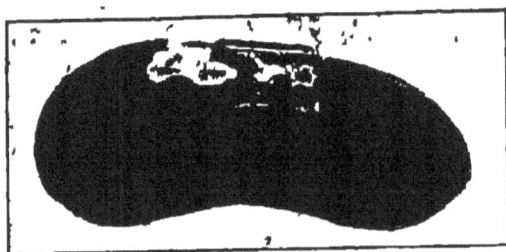

Fig 39 *Unio margaritifer*

appear to be at all hurtful beyond slightly impeding the progress of the fish to which they are attached in their movements through the water.

The *Unionidæ* are similar bivalves, not quite so large as the foregoing, but heavier in the shell, and they seem to thrive in rougher water than the *Anadonta*. They often furnish very fine pearls, especially the *Unio margaritifer*, which at one time was our principal source of pearl supply in this country.

The *Cycladidæ* and the *Pisidia* are small cockle-like bivalves, varying in size from three-quarters to one-sixteenth of an inch in diameter. Trout eat them greedily, and we well remember one large reservoir where the rapid growth and fine condition of the trout was ascribed to the abundance of this food supply, and the trout that were caught certainly had a great number in their stomachs. The food of all snails consists of vegetable growths and decaying matter. They are excellent scavengers, and destroy no trout food, so that the addition to the other inhabitants of a pond is a clear gain. With the exception of the *Paludina* they are all oviparous, and multiply very fast under favourable conditions. There is just one slight danger with regard to bivalves, and that is the fact that they feed with their tongues out; small fish occasionally seize hold of these and get their noses drawn into the grip of the shells. They have occasionally been seen swimming about with one of the *Cycladidæ* firmly adhering to them, having been stoutly gripped by the nose as a reward for curiosity. Occasionally, even unwary birds and animals fall victims, and on one occasion so large a bird as a water hen (*Gallinula choloropus*) was found dead from having attacked a swan mussel (*A. Cygnea*). The beak of the bird was firmly held as in a vice, and the huge bivalve being too heavy to carry about, the bird was simply drowned.

It is well known to most people that the water contains animals which are parasitic on fish. Probably few persons, however, are aware of the great variety of these creatures. Some are exceedingly obnoxious, whilst there are others again which seem to be almost harmless. The sea-going or anadromous fishes are attacked by parasites in the salt water, which they get rid of on coming into the rivers, and *vice versa*, but it is chiefly with the fresh water parasitic animals that attack trout that we have to

Fig. 40.   *Cyclas cornea*

deal in this chapter. There is one,
which attacks the young fish, known
as *Trichodnia pediculus*, a minute
creature just visible to the naked
eye. They are excellent free swimmers, and sometimes attach themselves to fry in large numbers, but

Fig. 41.   *Trichodnia pediculus* (en'ar'd)

no special harm seems to follow from their presence, so far as we
have observed. When viewed under the microscope they make
beautiful objects and have some resemblance to a sea anemone.
They are easily killed by placing the fish in a saline bath.

Another parasite that occurs on fry is probably one of the
Trematoda, which has the appearance of a small worm, and makes
another very beautiful and interesting object for microscopic
study. It is *Gyrodactylus elegans*, and is possessed of fourteen
cat-like claws, with which it clings to its host. It is viviparous,
and on several occasions we have witnessed the birth of a young
one, which at once finds a suitable *nidus* on the fish, and
commences to feed.

Fig 42. *Gyrodactylus elegans* (enla d)   Fig 43 Foot of *Gyrodactylus elegans* (enla d)

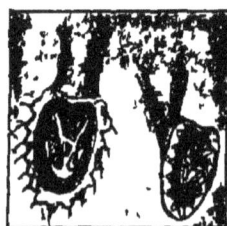

It has been suggested that there may be some connection
between these parasites and the fungoid growth (*Saprolegnia*)
which is often found on fish. After a most careful and extended
investigation, we could not, however, find facts to confirm the
slightest suggestion of any connection between the two. With
the aid of a powerful microscope we have watched them feeding

frequently, but whether they actually remove anything from the fish, or simply anchor themselves, and there feed on very minute *animalculæ*, we could never be quite sure; but this much is certain, they do not leave behind them a scar of any sort that we could see.

The members of the leech tribe *(Hirudinea)* are all more or less dangerous to fish. They leave scars which soon form suitable resting places for the spores of *Saprolegnia*, and this often proves fatal. The great flat-footed leech *(Geometra piscicola)*

Fig 44 *Geometra piscicola* (enlarged).

and its cousin, the horse leech *(Aulostomum gulo)* are both unwelcome visitors in fish ponds. The horse leech is very common in many waters, and is well known, and when it attacks fish, the loss of blood from its tri-radiate bite is very weakening. It is true that the trout will eat the leeches, but we much prefer keeping them out of the places where trout are, as they destroy more than they produce as food. In their youth, their principal food consists of the larvæ of flies and small insects. They seem particularly fond of caddis worms and soon unhouse and eat them—indeed they seem almost designed by Nature to be the *bête noir* of these creatures. All the leeches are free swimmers, progressing by an undulatory vertical movement. They also crawl fast, and are capable of living for some time out of water, so they are not very easy to get rid of. Fortunately they are not very prolific; about twenty eggs per annum seems to be their rate of increase.

Most anglers will have met at some time or other with a beautiful greenish crab-like creature adhering to fish, which is commonly called a "tick." They vary in size from one to three-eighths of an inch in diameter. The common one is *Argulus foliaceus.* We have found them on gudgeon frequently, and also on trout and pike. They swim freely and adhere to their host by means of powerful suckers, and leave a bare patch wherever they have adhered, thus rendering the fish peculiarly liable to an attack of fungus. A saline solution is the best remedy for this, and probably for most fresh water parasites.

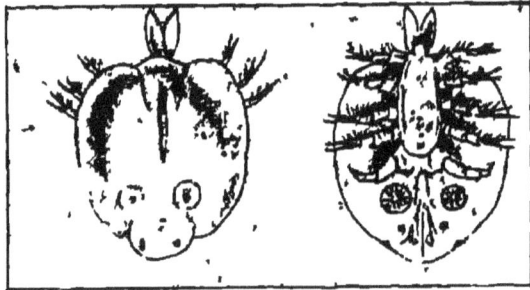

Fig. 45   *Argulus foliaceus,* enlarged.

Some of the worst parasites are the Entozoa or internal worms, which sometimes are found to be the cause of a considerable fatality amongst fish. We have taken a tape worm over two feet long out of a trout, as well as a dozen or so of small round-pointed worms, probably *Annularia.* These may be introduced with food as cysts, and then develop, but as they are built in segments, and every joint or section may become another worm if detached, it is highly desirable that no domestic animal be drowned in a fish pond, or in a stream leading to one. Dogs are frequently troubled with worms, and the drowning of a dog might easily be the source of introduction, eventually causing serious disaster amongst the fish inhabiting the water.

Every one probably has heard of the parasite commonly known by the name of "fungus," or "salmon disease." It is known to fish culturists as *Saprolegnia ferax,* but there are other species, probably which will attack fish. There is another *(Leptomitus clavatus)* which attacks dead fish and decaying meat, but the *Saprolegnia* will grow on either the living or the dead. It grows in the water, and it can be grown in air in a hermetically sealed vessel, but it will not grow in the open air. When the water is low in our rivers and fish lie long in the pools, it often happens that the "fungus" assumes an epidemic form. Therefore if we can increase the supply of water, by producing an artificial spate, the fish are helped to reach the sea, and as sea water has been ascertained to be fatal to it, the benefit is apparent.

Saprolegnia is a cryptogamic plant, belonging to the group *Thallophytes*, and it occupies a position in that group between *Siphoneæ* of the *Algæ* and *Phycomycetes* of the *Fungi*. At one time it was a disputed point whether it belonged to the animal or vegetable kingdom, but there is now, I think, no doubt about that. The only doubt that remains seems to be its exact position as a cryptogam. When seen upon a salmon or a trout it has a woolly appearance, and often occurs in patches the size of a shilling or a half-crown. Sometimes it covers the whole of the head of a fish, whilst in others it takes possession of the tail or of the back, or grows in a small tuft from one or more of the fins.

If we examine this woolly-looking mass through a strong magnifying glass, we shall find that it consists, first of all, of a matted-looking mass of filaments, lying comparatively flat upon the body of the fish. This practically constitutes the root, and is known as the *mycelium*. From this root, or *mycelium*, rise a multitude of filaments or stems, called *hyphæ*, each one of which consists of a hollow tube, and a great many of these tubes have their terminations enlarged, and assume a somewhat club-shaped form. This club-shaped vessel is a *zoosporangium*, and is found on examination to be full of minute seeds or *zoospores*. These, when ripe, are rapidly expelled from the *zoosporangium*, or seed vessel, and have the means of locomotion, moving about in the water by means of their two *cilia*, which may almost be likened to a pair of oars.

Until the plant is ripe, and begins to give off its seeds, there is not much danger of infection, and in its earlier stages it is easily destroyed by permanganate of potash or carbolic acid. To kill and bury every fish that can be got out of a river with any fungus upon it is a great mistake, and yet I have known this to be done. A great many of the fish so destroyed might have been saved, the fungus upon them not being at the time particularly

dangerous to other fish. I have seen fish eat pieces of fungus
without apparently receiving any harm, and on introducing a considerable growth of the plant into an aquarium, none of the fish
contained therein were attacked, nor did they suffer apparently
the slightest inconvenience from the presence of the pest.

That fish in a sound healthy state seem to be comparatively
safe from its attacks, under ordinary circumstances, seems to be
tolerably certain. What those extraordinary circumstances are,
that tend to render fish at times so liable to be attacked, require a
great deal of investigation. We know a good deal about the fungus
itself—indeed we are by this time well acquainted with its life
history—but we do not know all we might know about the salmon.
There is a "missing link" in the life history of that fish that
leaves us still in ignorance, and when our knowledge is sufficiently
increased, we may be able to apply some remedy for the great
scourge which has ravaged our rivers, as well as those of other
countries.

After trying a great many experiments upon fish of one kind
or another, I am satisfied that there are several distinct conditions
under which the fungus exists upon their bodies. The result of
my experiments leads me to the conclusion that, as a rule, fish
that are strong and vigorous in constitution, and at the same time
sound in body, are comparatively safe from the attacks of fungus.
But there are many modifications of this rule. I will refer to
some of them :—

1st.—Take a strong healthy fish that has got wounded on the body, or on a
fin. So long as the wound remains clean fungus will not necessarily
grow, but sooner or later little bits of flesh or skin become dead, and
decay sets in. Then the fungus grows upon them. This I have witnessed
repeatedly, and also that under such circumstances fungus will remain
for a long time without spreading. It is quite easy to remove it, and
while it exists the fish does not seem to be any the worse for its
presence.

2nd.—Fungus from some exceptional cause or other will attack the tail
or other part of the fish to a considerable extent, say a patch as large
as a half-crown. The rest of the body remains in an apparently
healthy state, and the patch of fungus goes on spreading over it from
the one centre only until the fish is killed, but no other fungoid growth
has been commenced upon it, and the uncovered portioned of the skin
apparently remains in perfect health.

3rd.—Fungus attacks a fish in patches, appearing in several places and spreading rapidly, the colours of the fish being perceptibly less bright, and the general appearance le.ding to the conclusion that it may be out of health.

There are some other typical forms of fungoid development, as, for instance, when it grows inside the mouth and on the gills; but I will at present only deal briefly with the three cases mentioned above. In the first case the entire body of the fish is in a healthy state, therefore the fungus, which will grow on a little bit of semi-detached dead skin or flesh, does not extend any further, and finally peels off. This often happens under natural conditions, and without any fish-cultural intervention.

In the second instance we have a healthy body, which from some cause suffers local derangement, as, for instance, from the effect of a bite or wound. An attack of fungus follows.

Here the local unhealthiness of the body, accelerated possibly by the fungoid development, extends, it may be, rapidly, and the fungus extends with it. I have seen cases of this kind in which the fungus has gone on spreading until nearly half the body of the fish has been covered, the flesh on that portion appearing as a mass of rottenness when exposed to view. I have put the fish under medical treatment, and have succeeded in stopping the progress of the fungus, and even in driving it back a little, but such a fish is usually too far gone, and dies under the treatment. The effect, however, is often to cause a cracking or breaking of the skin, and a partial separation of the diseased portion of the flesh. The process of sloughing in reality begins, but the mischief having gone too far the fish has not power left to throw off the diseased portion, but simply makes an attempt to do so and dies. In cases of a milder form which I have treated the sloughing process has been successfully completed, and the whole patch of disease, both fungus and skin, has peeled off, the wound has been carefully treated with carbolic acid, and the fish has recovered.

In the third case we have an unhealthy fish which is attacked, and, as the whole body is practically in a fairly good condition for the fungus, we find it growing in several places simultaneously. It sometimes takes but little to place a fish under such conditions. Wild fish, caught and placed in a small tank, for instance, and

Q

pining for their liberty, get into a morbid condition that renders them peculiarly liable to attacks of fungus.

I think I may safely say that I could cause fish to be attacked with fungus by simply placing them in a wooden tank, or in a newly-made one of concrete. I have done so, and have then removed and cured them. One very interesting and important feature I have noticed, and that is that a fish that was not itself attacked by fungus, on being removed transmitted it to others. I have taken fish that have been affected, as in case No. 3, and have destroyed every scrap of fungus upon them. In a few days a fresh crop has been found growing, and the fish has been put under treatment again, and again a third, and perhaps a fourth time. If taken during an early stage of growth, four out of five of such fish are curable.

The average life of a trout is about ten years. When dying of old age they are largely attacked by fungus, which kills them. On a fish farm they are not allowed to die of old age, but are killed and marketed before that time comes. I had rather an interesting case a few years ago. A Buttermere trout, which I had kept by way of experiment, died, at the age of over seventeen years. When in its prime it turned the scale at six pounds and three-quarters, but at the time of its decease it had become long and lanky, and weighed under five pounds. It was attacked by fungus, which was removed by means of a saline bath and the use of carbolic about a dozen times in as many weeks. The fish then succumbed. Its allotted time had come. This happens to most fish that are not caught, and the use of fungus seems to be to destroy not only the lives of these old fish, but to live on their bodies also, and it fulfils its mission. They are its legitimate prey, and are designed by Nature to be so.

In all cases of fungus on fish that have come under my notice, I think I may safely say that there has been a predisposing cause. Either old age, a morbid condition, wounds, removal of the mucous coating, or something else, has been underlying the attack. To preserve our fish from this deadly enemy as far as possible, our aim must be to keep them in health. If we allow them to deteriorate, and to assume a low state of vitality, we give the fungus a better chance. A great deal more depends upon the condition of our fishes than has been supposed, and one object of this work has been to show how we can deal, at least with some of them, and, by supplying the necessary conditions, how we can largely improve both the quality and the quantity, and make the waters yield their increase according to the designs of Nature, who has placed within the easy reach of man all the essentials for the making of " An Angler's Paradise."

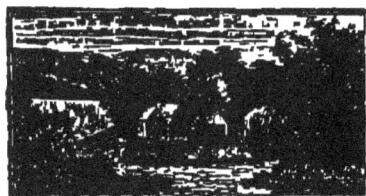

CHAPTER XII.

REARING THE FRY.

WE have now reached a very important point in the course of fish-cultural work—the time at which the little fish begin to take food. When they have nearly absorbed the sac, and for that reason have "unpacked" and scattered themselves over the boxes, they may be allowed a little light. A very good way of supplying this want is by making one set of lids do for two boxes, that is to say only half covering them. At the Solway Fishery the boxes are arranged in couples, side by side, and the lid is placed so as to half cover each of them. The lower half of the box, which was left uncovered during the "alevin" stage, is now half covered, like the rest.

A short while before the complete absorption of the umbilical sac a few of the fish will be seen to rise from the bottom of the hatching or rearing boxes, and for the first time take up their position, as if they had an object in view. They have their heads to the stream, and appear more like fish. If some particles of chopped egg, or indeed of almost any kind, be allowed to float down with the current, they will turn aside to seize them as they pass. Soon the number of fish acting thus will be largely increased, and from this time forward they require regular feeding and attention. There has been a good deal of controversy as to the best time at which to commence feeding young trout, some persons holding that there is no need for it until the umbilical sac is quite absorbed. Others, again, advocate their being fed a short time before the final absorption of the sac.

I think it may be taken for granted in the case of young trout, that when nature teaches them to look for food it is right that they should have some given to them. Like all other creatures, they have to learn to eat, and they do not make much "fist" at it at first. Indeed it would be somewhat surprising if they did. It will soon be evident to an observer that they will seize any particles floating past, no matter what they may consist of, and most of these, instead of being swallowed, are ejected again, and this by one fish after another as they drift along the current. The little fish have not learnt to know what is food and what is not, and at first they will snap at anything and everything that they see in the water, provided it is moving.

It is rather a critical period of the trout's life, for though there should be very few if any deaths amongst them at this stage, yet there may be heavy loss later on if they be mismanaged when just commencing to feed. Now is the time to train them to eat the food upon which it is intended to rear them. I have seen them a few weeks later in life refuse some of the choicest food that could be provided for them, simply because they had been trained to eat something else, and had got accustomed to it. They can be trained at this period to eat almost anything. The question very naturally arises, What is best for them?

I believe I am right in saying that the best food for young trout has not been discovered yet; that is to say, the means of procuring it in sufficient quantity to provide for the wants of a large quantity of them. If we open the stomachs of a number of wild trout fry we find them filled with various *Entomostraca* and other minute creatures which inhabit our waters. In a state of nature they can usually obtain a supply of these. In a stream, for instance, where we find a young trout here and there a yard or so apart there is a chance of sufficient food turning up to keep the little fish going. They are continually on the look-out for it, and each individual fish takes up its station in some suitable place where there is a current or an eddy, where it is comparatively safe from its enemies, and yet where an abundance of food is brought to it by the current. In a hatchery tank, where there are some four or five fish to every square inch of surface, the matter is widely different. If these are to be fed on natural food, it is clear that it must be produced for them, and that in large quantities.

In order to produce crops of natural food, accessory ponds are required, and much may be done also by growing a large quantity of *Entomostraca*, etc., in a pond, before turning in young fry, so that on the absorption of the sac they may find the best means of subsistence close at hand. This is a subject on which a chapter may be written, and I hope some time to go more fully into it. There is a great diversity of opinion as to the best time for turning out trout fry. It was the custom a few years ago to retain them until they had got well on the feed, whereas now it is becoming a common practice to turn them out just before they begin to take food, that is just before the final absorption of the umbilical sac. Taking a general view of the subject, I am in favour of the latter plan, except in those fish-cultural establishments where everything is perfectly arranged, and the work thoroughly understood by skilled workers, and where the fish are to be reared artificially to yearling or larger size. Where fish are simply grown for the purpose of keeping up the supply in a lake, pond, or river, and are to be turned out as fry, I believe thoroughly in turning them out before feeding into prepared raceways and ponds where that can be done ; otherwise into tributaries of the lake or river. Such a course saves a good deal of trouble, and I have seen excellent results obtained by following it. The little fish do best in a small tributary or raceway where there are plenty of water plants. A watercress bed is an excellent place.

Should it be decided to retain the fish in the boxes, they must be cared for at once when they begin to look for food. The best thing to do is to place a tank well filled with *Entomostraca* in such a position that the water may be run into the rearing boxes. Should there not be sufficient of this natural food available, a supply of the artificial substitute must be provided. The simplest article to use, and one which is easily obtainable, is raw liver, which must be grated or chopped very fine, and the coarser particles taken out by screening through some very fine woven fabric or fine perforated zinc. I have found the latter answer very well. Another very good plan is to separate the particles by decanting. To do this, a glass jar such as one sees in the window of a confectioner's shop is useful. It possesses the great advantage of transparency, so that an operator may see what he is doing, otherwise it is very easy to pour out a quantity of the settlings which should be carefully avoided.

Having selected a jar that will hold about a gallon, put a pound of grated liver into it and fill with water, then stir until well mixed. Allow a moment or two for the coarser particles to settle, and pour off the liquid as far as can be done without letting the larger pieces go over. Then add more water to that which is left in the jar, and stir again. A second lot may now be poured off, and the remainder may be thrown to the yearling fish, or, if desired, chopped up again until another lot can be taken out of it for the fry. The liquid that has been decanted may now, with the help of a funnel, be put into pint bottles, and from these fed to the young fish. The process is quite easy, all that is necessary at first being to pour a quantity of the liquid into each rearing box at the head where the water comes in, and watch the result. In the first place the liquid is thoroughly disintegrated by the in-flowing current, and the minute particles of food are hurried along by the current, and the fry will soon be observed to be taking them.

If the fish be well up towards the head of the box it may not be necessary to feed elsewhere, but should they be very much scattered through its length some of the liquid should be poured in lower down. By all means take care that all the fish have a chance of getting some of this food, if they have any desire for it. It is a great point to get the fish well up to the head of a box whenever possible, but sometimes it is not easy. If they will not come to the food, the food must go to them, or the result will be serious. A great deal of care is required in the early feeding, as much depends upon it. A novice will find a good deal of practice and skill is required, and also that there is a good field for observation, and that much may be learnt that way. An expert will go through a large hatchery, bottle in hand, and feed the fish in a large number of boxes in a comparatively short time, taking a number of boxes in hand at once, and going over them several times. Several machines have been invented for feeding fry, some of which are ingenious. The principle is a rotary wheel, which is charged with food, and is so' arranged as to be constantly letting out a small quantity in the course of its revolutions. These machines have not, however, come into general use. Fish do not feed so regularly as the machines would have them do. Sometimes they eat more and sometimes less, according to surrounding circumstances, such as temperature, light, atmospheric conditions, etc., and a fish culturist is aware of these changes, and can act accordingly. I have known a machine go on discharging food into a rearing box when the fish were off the feed, until the water became sickly, and it does not take long to bring about such a result.

There is a great advantage, too, in having the boxes and fish under constant supervision, which is the case if hand-fed. A caretaker can see any little change or notice any irregularity that may occur anywhere, and it can have his attention at once. It often happens that such attention is necessary, and will prevent great injury being done. or perhaps great loss of life. The fish should be carefully watched and their every movement noticed from day to day, comparing one day with another, and in this way much will soon be learned that cannot be acquired from any books. The food should be varied a good deal, and I use a large quantity of eggs for feeding the young fry. They may be used either raw or boiled, and, if raw, the yolk and white may be mixed together and water added, and the liquid decanted as described in the case of liver. Curd made of rennet and milk is also excellent in limited quantity, but both this and egg should be cautiously used. Too much of either may prove hurtful.

The best artificial food that I have ever met with, however, is shrimp paste, and this I have used as well as mussel paste, for feeding and rearing fry, and have found it most successful. To make it the shrimps have to be boiled and shelled, and the same applies to the mussels when they are used. A No. 22 " Excelsior" chopper, which may be obtained through any ironmonger, will be found excellent for preparing this or other food for fish of any description, small or large. I have had an extra fine plate made for preparing this paste, which may be procured·with the machine from Mr. W. Barnes, ironmonger, Ashbourne, Derbyshire. The food is expensive, but the results are good, and fish reared upon it will do credit to the producer, and are worth more than fish reared upon anything else I have met with. I am often asked how many times a day fry should be fed. I have tried the plan of feeding them every hour, and also of feeding them four times a day, and I incline to the latter. If fed too often they are apt to get into a lazy way. When fed four times a day they get a good appetite between meals, which makes them eager for their food, and they come for it better than they do if fed oftener.

When the fry are well on the feed and have got thoroughly used to the system, they should be transferred to the rearing ponds. At first they will exhibit symptoms of alarm when the attendant comes alongside the boxes, and will all dart away from him, but soon they get over this, and begin to associate his appearance with something to eat, and it is noticeable that they look for this something and expect it to come, and when it does come they "go for" it at once, and it is quite a pretty sight to see them feeding. Now is the time for the transfer. The ponds should be in readiness some time previously, with screens fixed and sluices working, and everything in order for the reception of the little fish. The first thing is to get them out of the hatching boxes. There are several ways. One is to lift the box so that the outlet end rests on a tub which is filled with water, taking care that the plug is over the tub. Remove it and the fry will be drawn off, the extra water escaping through an aperture covered with perforated zinc in the side of the tub. The upper end of the box should be raised slightly, and as the water goes down a current should be kept up over the bottom to prevent any fish being left dry. This may be done by means of a bucket of water or from a hose pipe. The latter is the most convenient where practicable.

Another method is to remove the screen and let the water and fry all escape through the outlet spout, finally turning the box on its side to get rid of the last of them. The tub is then slung on to a carriage and wheeled away to the ponds. The plan which I adopt, however, and which I find the best, is to syphon out the water and the fry, and with a little practice it can be done quite easily and without doing any harm. Work the syphon at one end of the box and let it discharge into a large pail or tub, which should be raised nearly to the level of the hatching box, so that the outlet end of the syphon can be kept always under water. There is then no sudden jar or shock, but the fish is simply drawn over by a strong current, which gradually disperses in the lower vessel. I have moved many millions of fry in this manner without hurting a fish.

From the hatchery they may be carried by hand in large pails, or in a tub slung on a specially made carriage. The distance at which the nursery ponds are located has, of course, much to do in deciding upon the mode of conveyance. On arriving there the fish should be very carefully turned out. A great deal depends on the care with which this is done. An easy method is to sink the pails, with the little fish in them, in the water of their future home, then to turn them gently over and withdraw them bottom first, floating or swimming the little fish out in the operation. However carefully it may be done the little creatures will get a great fright, and a large percentage of them will be a good while in getting over it. Finding themselves, after all the agitation caused by their transfer, in a strange place and in deep water, many of them will go to the bottom and remain there for twenty-four hours or more. I have seen some of them as long as three days in this position at times. This is very bad for them, and many of them never get over it. Others, again, will scatter all over the pond, which is also undesirable, and should be prevented as far as possible.

The old plan just described answered very well, however, on the whole, when the arrangements were well carried out and much care and patience bestowed on the little fish after being placed in the nurseries. It often proved a tedious matter, though, to get them on the feed again, and also to get them up to the head of the pond, which is the best place for them. As a rule they will in any case divide into two bodies or shoals, one occupying the head of the pond where the water comes in, and the other the end where it runs out. The skill of the attendant is here brought into play, his object being to coax the fish at the lower end of the pond, by every means in his power, to come forward and take the

SOME NURSERY PONDS, SOLWAY FISHERY.

.

higher place. They should gradually do this, the best fish coming out of the shoal each day and pushing up to the head of the pond. When this is happening it is an encouraging sign, and as the shoal is decreased in size the one at the upper end of the pond should correspondingly increase. At last there will only be a few weakly fish left near the outlet screen, and when it is seen that the fish have well sorted themselves, it is a good plan to take these fish out and put them into a pond by themselves, or into a small stream. They will have a better chance, and some of them will make good fish, but they will not do well if left in the nursery pond, unless they get a good deal of extra attention.

Whilst the fish are being coaxed up from the lower end of the pond great care and much watchfulness are required, lest those occupying the head waters should begin dropping down to the lower end. Sometimes this will happen in spite of all the attention that can be given them ; indeed, if there be many fry to look after, the entire time of one man will be required for awhile, and it will take all the skill and energy he can bring into play to bring the little fish safely through the difficulties that beset them at this period of their existence. From morning till night his attention must be given to the ponds. Another man prepares the food and brings it to him in jars, and the contents of these jars are placed in the feeding boxes ready for use. These feeding boxes consist merely of perforated zinc boxes or cages, fitted with handles about a yard long. The box containing the food is dipped into the water, which immediately enters it by means of the perforations. On being lifted the water runs out again, carrying with it such portions of the food as will pass through the perforated zinc. When done, the coarser particles may be put through the chopper again, or they may be fed to the yearling fish, as may be most convenient.

The most successful plan for turning out young fry, and one which I have adopted for some years, is by means of floating boxes, and its success has been far beyond that of the old system. Seeing how frightened the little fish were on being transferred to the ponds, and how diffic. it was to coax them up again and get them into good order, I tried the plan of floating a hatching box in the pond, and turning them into that instead of into the pond itself. The box may be arranged to stand on legs, or may be supported by two ropes passed underneath it and fastened to pegs stuck in the ground on each side of the pond. It should be placed at such a level that a sufficient current may be directed through it. Where there is a fall averaging some twelve inches between each pond, as at the Solway Fishery, the matter is exceedingly simple. Where the fall is necessarily very slight, owing to surrounding levels, etc., it is best to have an independent supply of water, conveyed in a temporary spout or by means of a hosepipe.

The boxes being placed in position, and the water turned on to them, the fry may be put into them. After their transition from the hatchery they are naturally frightened, but instead of being scattered in a pond that is strange to them, they find themselves again in a hatching box, and none the worse for their journey. The fact of their being altogether is also reassuring to them. They cannot run away, and the little excitement they have had has, apart from their fright, given them an appetite. In half an hour some of them have begun to make themselves at home, and a little food may be given to them. It should only be introduced at the head of the box, and the few fish that are ready for it will encourage others, who see them feed and are themselves hungry. They will gradually come up from the bottom of the hatching box, and in a short time all will be feeding and apparently quite at home again. It must be a delightful discovery to find that after being syphoned out of their original home, jolted in a tub or a can, and then poured out of it, that they are not killed or hurt, and that after all they are only in another box which looks very much like the former one, a place where supplies of food are again given to them. Some of them feel hungry, and take it, and find it the same agreeable mixture that they have been accustomed to. Others that are looking on begin to think they would like some too, and the rest follow their example, and soon all are enjoying an ample meal. With regard to the past their fears vanish, and they begin to think they have had a dream. They should be treated while in this box exactly as they have been dealt with in the hatchery.

The next day they will be happy again, and feeding

vigorously; and, this being the case, the screen at the outlet end of the box may be carefully drawn, so as to allow them to escape at will. This should be done very gently, and quite unknown to the fish. The best time to do it is just at the commencement of a feed, when their attention is attracted by the food which is being given to them, and the feeding should go on just as usual. When it is over the fish will fall back a little, and some of them, without knowing it, pass out of the box. There is not the least fear now, and instead of gravitating to the bottom of the pond, and lying there in a half-terrified state, there is a delighted feeling of curiosity and pleasure at being free and in such a spacious apartment The current of the inflowing stream is felt, and the little trout, for such he now truly is, works his way up to the head of the pond, on the way snapping up a specimen of *Cyclops quadricornis*, which he finds delicious, and he soon takes up his position looking for more. This goes on until fish by fish the box is vacated, and when all have left it may be carefully removed. This should be done without disturbing the fish, and it may be desirable to slaken the ropes a little and float it down the pond before removing it.

It often happens that a few fish take up their position in the box and will not come out of it. Should such be the case, they may be ejected gently by turning it over on its side before removing it. They are fish that will very soon take care of themselves when in the pond, and no anxiety need be felt on their account. After the fry are out of the boxes the feeding should go on carefully, and special watchfulness should be exercised at first until they are all thoroughly on the feed, and in skilful hands this is not long.

Having got the fish into good feeding order, the great object to be desired is to keep them in position, and not to allow them to scatter over the ponds. Great care is necessary just at this crisis and for some time afterwards, that the fish are not disturbed.

A very little thing will sometimes suffice to throw them off the feed at this time. A heron flying over, or some visitors appearing suddenly amongst the ponds, a thunderstorm or a sudden spate, will often effectually prevent fry from feeding for awhile.

Everything shouuld be done, therefore, that is practicable to ensure the ponds being kept quiet. Covers should be provided, under which the little fish can take shelter from enemies, real or imaginary, and from the rays of the sun when desired. The ends of the ponds where the fry most collect should be covered with netting at night to make them doubly secure, or a sharp look-out should be kept for herons, kingfishers, etc. It is better for awhile that the fish should see no one but their attendant. They soon get accustomed to him, but will sometimes scatter at the sight of a stranger, and require a good deal of coaxing to come together again. With a good flow of water and pleasant surroundings for the fish, in the shape of marginal plants, they should thrive well. The plants will tend to provide a supply of natural food, and are most useful adjuncts to the rearing ponds. The little fish will take up positions under their leaves and about their roots, and it will be found that those which do this will make the best fish.

Some years ago I had the grass kept carefully cut between the ponds. A well kept sward looks very nice, but if allowed to grow, the grasses and plants produce a considerable amount of insect life, which is worth far more to the fish than a well kept lawn. I would not say one word against keeping the banks of the ponds cut close, but just give the result of my own experience, and must leave others to judge for themselves and to do as may seem best. The long grass and other herbage is a temptation to water rats, but these can be kept down. Sometimes the fish jump out on to the bank, and falling amongst long grass, cannot get back into the ponds again, and die. Whilst the grass is uncut, however, a constant growth of insect life is kept up, which is valuable, and the grass when cut, as cut it must be when fully grown, is useful either in its green state or for hay. It is a good plan to cut the grass on one side of the pond only, and to allow it to grow up a little again before cutting that on the opposite bank.

I am often asked—How much food ought to be given to a pond full of young trout? In answer, I would say that it is impossible to lay down any definite rule ; some lots of fish will be found to feed more freely than others, some ponds will contain more natural food, and so many little things come in between the feeding of the fish and their attendant that much must be learned by experience; the knowledge certainly cannot be gained from books. When the fish come up for their food with a good appetite, as they will if left long enough between meals, it is a good sign, and they may be freely fed. The distribution of the food should be managed carefully, and not too much given at one dip of the feeding box. Much care is required that no food is allowed to go to the bottom uneaten. Should it be allowed by any accident to get there, it must be carefully removed, which may be done by means of a fine gauze net. Sometimes, when the feeding of fry is managed by beginners, the bottom of a pond is allowed to get covered with uneaten food. I have seen such ·cases, and what has happened may happen again. Should such an occurrence as this take place it would be very difficult to remove all the food, and if nothing be done it will assuredly pollute the water, and probably produce disease, which may soon play sad havoc amongst the fish.

The best thing to do in such a case is to scatter a little finely-sifted earth over the pond. It will settle on and about the offensive matter, and if this be followed up by a good covering of clean gravel, the bottom will once more be rendered clean and pure. Earth is an excellent thing to use occasionally in the ponds. It is good for the fish, and is such a powerful deodoriser that it tends to absorb, and so keep down any impurities that may exist. It is often desirable to use it in the rearing boxes before the fish are turned out of the hatchery. Some good clean earth should be selected. A few sods from a good old pasture are the best, and, to use them, they should be put into a tub full of water and well stirred and shaken, so that the earth is washed out of them. Then the liquid may be poured off into the pond, or the inside boxes, as the case may be. Pour it off carefully, and if screened through fine perforated zinc so much the better, as this prevents lumps and fibrous or rooty matter going into the pond. It is only the muddy water which is required, and there need be no fear in using it freely.

I have often watched the effect of doses of earth given in this way, and have found them to be highly beneficial. In a natural stream a considerable quantity of earth comes down with the water every time there is a freshet, and in the rearing boxes and in nursery ponds this is not so, therefore the want must be supplied artificially. Every one who knows anything about trout knows how they delight in a rise of the water, and how much more easily they are caught at such a time. An examination of the contents of a few of their stomachs will show, in addition to the ordinary food, consisting of worms, flies, etc., a proportion of gravel and earthy matter. It is sometimes very desirable to make artificial spates in fish ponds, which is easily done by damming back the water and letting it off through the ponds. Care should be taken that the flow through the ponds is not materially lessened in performing this operation.

Earth has a beneficial effect in many ways, and in some cases has been found to be a cure for fungus. Fry are very liable to be attacked by fungus, and too much care cannot be exercised in endeavouring to keep clear of it. Avoid any uncharred wood-work about the ponds. Be careful that no roots of trees that may have found their way into the raceway are cut in the spring. I once saw a raceway that had been cleaned and had its sides trimmed in early spring, and in doing this the roots of several trees had been cut away. In a short time the end of every stump was covered with fungus. It is also largely produced on the pond bottoms when they become foul. Therefore, keep them clean, for if they be allowed to remain dirty mischief is sure to accrue to the fish. One of the most fertile sources of danger is to be found in the accumulation of greasy or fatty matter on the surface of the water, the result of artificial feeding. It should be got away as often as it accumulates. Every time the fish are artificially fed, more or less of this grease must be let free, and, if not attended to it will float on the surface of the water. The little fish, when swimming near the surface, will come in contact with it, and it will adhere to their fins and bodies, and will become a suitable *nidus* for the germs of the fungus. It can easily be kept from accumulating by breaking up the scum which it forms, and floating it on to the outlet screen, or by skimming it off the surface of the water.

Should fungus from any cause attack the fish salt is a remedy, and it may be applied either by putting it through the ponds, or by netting out the affected fish and giving them a salt bath. In the latter case the fish should not be returned to their pond, but be placed in another small one by themselves, so that the dose can be repeated where the first is not efficacious. It should be borne in mind, in putting salt through the ponds, that it is destructive to some forms of life, and there is a danger of its disarranging the equilibrium of life in the water, which has an important bearing on the welfare of the little fish. The salt bath after removal from the ponds is the best remedy. The solution may vary in strength. The weaker it is the longer the fish can remain in it, and the stronger it is the sooner they require removal. I prefer a moderately strong dose of salt, though by no means in excess, but different practitioners have such varied opinions as to the strength of the liquid required, that I would advise the beginner to try a few experiments with a small number of fish at first, and then decide as to the best course. A very strong dose destroys the fungus, but time is required, and if the fish turn sickly at once and have to be taken out, the liquid may not have had time to saturate the fungoid coating, and so the desired object will not be attained. A milder dose is often more efficacious, the fish remaining longer in it. The signal for taking them out is their coming to the surface or going over on their sides, when they should be at once removed to a current of good water, which will soon revive most or all of them. Of course a few will die, as may be reasonably expected in the case of such delicate little creatures, but in skilful hands the death-rate should be comparatively small, and sometimes it is almost *nil*. Sea water is better than salt and water where it can be readily obtained.

As the season advances, and the little fish grow, the quantity of food will require increasing, and when the death-rate is low they will soon require thinning out. For this purpose it is necessary to have some extra ponds, and they may be a good deal larger than the nurseries. A pond about sixty feet long and sixteen or twenty feet wide, with water four feet deep, will do very well, and if in autumn the level can be raised to six feet so much the better. If the fish are not thinned out they are very liable to contract disease and die. Some of them that have grown faster than the rest will also become cannibals and devour their fellows. The pollution of the water by being breathed over by so

R

many fish is considerable, and is a thing that is not often taken into account by practical men. It is a matter that is of vital importance, however, as the fish, although they may live, will be liable to be killed by circumstances which would not otherwise affect them. The slackening of the water supply—accidental or otherwise—a rush of surface water after heavy rain, or even a freshet, may kill a lot of them, and often the best fish are those that suffer under such circumstances. Do not overcrowd the fish therefore.

A pond sixty feet long, four feet wide, and about three feet deep, will hold ten or fifteen thousand fry at first, and give them plenty of room to grow, but by the end of July the number should be reduced to five thousand, which may be left till October, when they should again be thinned out, or better still, put into larger ponds. When pressed for pond room I have put as many as thirty thousand fry into a nursery pond of the size just mentioned, but they soon required thinning out, and whilst they were in the pond had a good current of water running over them. It is most important that the water supply should be at all times ample. There should be more than is actually required, so that at any time an extra quantity can be turned on. The amount of water run through each set of nursery ponds should depend a good deal on surrounding circumstances, but from fifty to a hundred gallons per minute is a fair quantity, and this will keep seven ponds going, other things being equal. If more ponds be required, then an entirely additional supply of water will be needed. At the Solway Fishery there are several sets of nursery ponds, and the water from each set is conducted into a raceway, and after flowing with a good ripple for a quarter of a mile is again used for supplying ponds containing larger fish.

A given supply of water will only support a limited number of fish, and after being passed through a series of ponds it requires purification. There is no better plan than allowing it to ripple over the stones, and to pass through a pond or even a raceway containing vegetation.

In this way oxygen is absorbed from the air, and is also received from the plants, which, on the other hand, take up the deleterious matter with which the water has become more or less

SOME NURSERY PONDS, MANAGER'S HOUSE, AND YEARLING HOUSE, SOLWAY FISHERY.

charged, and it becomes available once more for fish-cultural purposes. Much may also be done for the improvement of the
water, by encouraging the growth of suitable plants in and about
the margins of the ponds, and in the raceways between them.
Having all in thorough working order, there is little to fear as the
season goes on. The most critical time with a young fish is
about a month after feeding commences. At this stage a considerable loss often takes place, and it may arise from several causes.
The food given has often much to do with it. It will be apparent
to anyone that such delicate little creatures as young trout, when
fed on artificial food, may very easily get their stomachs disordered.
This undoubtedly often happens, and a great mortality is the result.

Some thirty years ago I found this out by losing a lot of
young trout fed almost entirely on boiled yolk of egg. They
were as nice a looking lot of fish as I ever saw, were feeding well,
and grew up to a certain point, when they sickened and began to
die off wholesale. In a week I had lost four-fifths of them. I
attributed the loss entirely to the food, and although now I use a
large quantity of eggs each season for feeding very young fry, yet
by judiciously varying the food the mischief is avoided. During
the first few weeks they require very careful handling indeed, and
their future depends upon the way in which they are managed at
this time.

Some food is too rich for them in quality, and this is the case
with yolk of egg. I fed another lot of fish entirely on chopped
worms, and they did very well, the food being much more natural,
and the amount of earth and grit which it contained being highly
beneficial to them. Over-feeding with too rich food is, therefore,
to be avoided, and the more natural food that can be given the
better. Sometimes, when a few weeks on the feed, some of
the fish may be seen to be getting very thin and "lanky"
looking, with heads large in proportion to their bodies. This is a
sign that something is wrong. It may arise from starvation, but
that starvation may be brought about in the first place by in-
judicious feeding. These fish will probably die, although if not
too far gone they may be cured by placing them in a raceway
where there is a good ripple and plenty of natural food. This
will cure them when nothing else will.

Sometimes, instead of appearing emaciated, the fish which look quite healthy and well will suddenly commence drifting on to the screen in large numbers. This is probably because their food has disagreed with them, and feeling sickly they are carried down by the stream, and will die if not removed. The best way of dealing with such fish is to place them, like their "lanky" brethren, in a suitable raceway, where many of them will recover. A raceway used for this purpose should terminate, if possible, in a pond, in which the fish can remain, and which should contain a good supply of natural food. They will remain long enough in the raceway to receive much benefit, and on reaching the pond will be stronger, and the change of diet will soon put them right again.

Many are the diseases to which trout are liable. A very sudden change of temperature will cause them to suffer from inflammation of the gills, which is sometimes fatal; often they get over an attack of this disorder with the result, some people think, of deficient gill covers. This may be a consequence, but I rather think that fungus is much more to blame for it. The fungus is apt to grow on the edge of the gill covers, and although it may be afterwards cured, yet it interferes with the development of the opercula, which become permanently contracted in consequence. Young trout sometimes suffer from parasites, and when such is found to be the case a salt bath is beneficial, and will usually put matters right, though sometimes several applications are necessary before the desired end is obtained. I have occasionally seen several cases of blindness amongst a lot of trout. It attacks the best and finest fish in the pond, and they become very light in colour. The strangest part of the malady is that they do not, at least for a while, fall off in condition. They are easily picked out at a glance from amongst the other fish, and are not worth keeping. When old fish become blind it has the effect of making them dark in colour and inferior in condition. Probably this is because they do not so readily adapt themselves to circumstances as younger fish do. I have seen the latter feeding freely, and evidently guided to their food by the senses of touch and smell, and they get very expert at finding food without taking much exercise. This accounts for their keeping so long in good condition.

In order to avoid the chance of disease amongst young trout
as much as possible, care should be taken to have good healthy
eggs from well-selected fish, and not only this, but eggs which have
been properly incubated in a well-appointed hatchery. Too much
stress cannot be laid upon these points, for with weakly eggs taken
from weakly ill-fed fish the grower has but a poor chance. There
is a great art in feeding breeding trout, which is only acquired by
long experience. To give them neither too much nor too little,
and to give them just the right kinds of food, at just the right
seasons, is a subject on which a book might be written. No book,
however, will ever teach the uninitiated how to do the thing. It
must be learned, as must many other things, by practice and a
thorough training at some good fish-cultural establishment. When
the trout have safely passed the crisis which occurs when artificially
fed, that is a few weeks after commencing to feed, and which we
call getting over the "distemper," the work becomes one of constant attention on the part of the attendant. The feeding of so
many mouths, together with the cleaning of the screens and the
regulation of the water supplies, takes up all his time ; and it is
most important that nothing should happen to the water supply in
any way. Should it accidentally be stopped, even for a short time,
great loss may ensue. Too much care cannot be taken so to
regulate the intake that it cannot err. The choking of a screen
may cause a pond to run over, which should also be carefully
guarded against, and ought never to happen amongst a good set
of ponds. Should all the necessary details be duly attended to,
there is every prospect of a good turn out of yearlings, notwithstanding the delicate nature of the creatures we are dealing with.
As they get older the danger of losing them rapidly decreases, and
by August they should be comparatively safe, having by that time
survived the dangers which fry are heirs to, and become yearlings.
It is true they are not a year old, but a great change has come
over them that entitles them to the name, which will be more
fully explained in my next chapter.

CHAPTER XIII.

THE YEARLING STAGE.

Salmonidæ adapted to cultivation—Rising to the fly—Fish culture requires experience—The food of yearlings—Must be properly dispensed—Development and selection of stock fish—Deformities—Pedigree stock—Sorting—Transit of yearlings —Netting— Preparation necessary — Caution to purchasers —Yearling nets— Yearlings hold their own against large trout—Two year olds.

IT is now a fact beyond dispute that the various members of the *Salmonidæ* are peculiarly adapted to cultivation. This applies both to the anadromous or sea-going fish, and to those which spend their lives in the fresh water. The latter section of the family can be dealt with by individuals, and this has been largely and successfully done. Much better results even can be obtained from the migratory section, including the salmon (*S. salar*), when the work is properly set about, but at present the amount of lethargy which is shewn with regard to this important matter is surprising. Trout have been materially improved, not only in numbers, which have yielded a heavy increase, but also in quality, which is better. Indeed, we have yet to find out what cannot be done with trout. By means of judicious treatment inferior races have been made into splendid fish, and now is the time to train them to take any special kind of food that may be desired. When in the fry stage they partake freely of food that they have been brought up to, but will often reject other kinds, and it is undesirable at that time to check their feeding operations. Therefore the food that is to be used continually should be the one commenced with. In July or early in August the food may be changed if desired, that is if any more convenient substitute can be found.

Now is the time to train them for rising to the fly, a most

important part of the fish culturist's art, and one that seems to have been entirely overlooked in the past. It was found out years ago at the Solway Fishery that by allowing the grass and other plants to grow, instead of cutting them too much, a host of live stock was produced, and the fish became accustomed to this from their infancy. It is a pretty sight, indeed, to watch the little fish in May, when just commencing life in earnest, rising freely to the fly. It is clear that when the sward is kept bare and the flies are not to be found, the fish cannot take them, and if they do not get the chance they lose the lesson, and a most important one it is. Where flies are produced in large numbers the little fish learn to take them beautifully; nay, they look for them daily, and depend upon them, and the lesson which is learnt in infancy, and impressed upon them in so practical a manner, is never forgotten. They get a liking for entomology which they never lose, and in after-life will do credit to their race.

There is undoubtedly a good deal to say in favour of inherited instinct, and a race of fish brought up to bottom feeding may be followed by a bottom feeding progeny. Even should this be the case, however, early training will prove successful, and the trained fish will win the prize against others. But by feeding them on the surface, and keeping a large number of fish in a pond, an eagerness is acquired to " go for " their food that is not known amongst wild fish. This tendency provides a wonderful facility for training them, and, as I have already said, they may be taught to do almost anything reasonable, if placed in the hands of an expert who is well up in his work. It must not for a moment be supposed that Dick, Tom, or Harry, if entrusted with the care of a lot of trout, would manage them successfully, even with the aid of books. It requires much patience and skill, and a, sort of "inherited instinct" or love of nature on the part of the man, without which the case is hopeless. The individual who possesses this faculty, and there are many who do, may succeed with due care and a sufficient amount of training.

A good deal has been said about fish, in these days of progress, ceasing in some places to rise as freely as formerly to the fly. The fact is that many waters are so overfished that the fly-taking trout get caught, and so gradually leave behind a bottom-feeding race. That this is really the case I can well conceive, from the experience I have had with trout in many ways. There is a remedy for it, however. I am quite convinced of this, although the application requires a little working out yet. We know quite sufficient to be assured of the fact, and have already learned a good deal regarding the training of fish. A vast deal more may be looked for from the fish culture of the future than from that of the past. Many difficulties that existed a few years ago have been overcome, and barriers to progress that at one time appeared insurmountable have, by patient endurance and repeated and untiring effort, been driven to the vanishing point.

Fish culture has been sneered at by some who failed in their first attempts, and cases have occurred where, through ignorance or careless working, the great results looked for and perhaps promised by the enthusiast, have not been realised, but this is only what might have been expected. I have repeatedly seen cases in which individuals who have perhaps read some fish-cultural work, or found an interesting paper on the subject in one of our magazines, have at once come to the conclusion that fish could be produced by the million by the expenditure of a trifling amount. It is needless to say that the work of such enthusiasts, entered into very often even without a single consultation with an expert, often results in failure and brings discredit upon a good cause. It is a well-known truism that where one man can live well another will starve, and this certainly applies to fish culture.

The idea that the making of a pond and the turning in of a lot of fish is all that is necessary must be exploded, and the work begun on right lines and under careful supervision. It will then be likely to produce results that will be gratifying to the promoter, and which will inspire confidence in the looker-on. To work out all the ramifications of the subject must be left to a few only, who are willing to make it their life work, but on the results of their labour may be built up a mass of information which will enable good work to be done where formerly it was unthought of. With a view to the improvement and promotion of the work, the Solway Fishery has been thrown open as a school of fish culture, and already this step is bearing good fruit. How gladly some of us a few years ago would have availed ourselves of such an opportunity had it been possible.

The natural food of yearlings, as indeed of fry, may be largely augmented by the introduction of various insects in their larval forms, not only as regards those which inhabit the water, but also the soil and its vegetation, and a study of flora and fauna therefore becomes very desirable. This applies not only to rearing or nursery ponds but to streams and lakes, which may often be very materially improved by the introduction and culture of certain forms of life which were formerly non-existent therein. As applied to our natural waters this subject becomes, then, an exceedingly important economic question, which is well worthy of the study and thought which is being given to it. Some of our winged insects simply abound in their earlier stages in certain localities, and under peculiarly favourable conditions. To supply these conditions in contiguity to our fish ponds is a work of great importance, but which has as yet hardly been entered upon.

Whilst bearing in mind the value of winged food, the more substantial and solid organisms should not be overlooked, such as *mollusca* and *crustaceans*. These play a most important part in the development of fine fish, and also tend to the introduction of that delicate pink tinge of flesh which is so much admired, and to a richness of flavour which is unmistakeable. That the food of fishes has much to do with the flavour of their flesh, fish culturists know quite well. It affects them just as much as the flesh of animals is affected by their food, and experience teaches that a variety of it is the best. It is natural that any animal should be benefited by a mixed diet, and that it is the case with the *Salmonidæ*, especially when they get older, there is no room for doubt.

Where large quantities of fish are dealt with it is desirable to use a certain quantity of artificial food. In conjunction with a good amount of natural food the artificial supplies are beneficial, and with due care in its distribution and a good flow of water the fish at this stage should give very little cause for anxiety. The daily routine of feeding them as their meal times come round, clearing the screens, and attending to the water supplies, is the constant work of their attendant, and very interesting work it is to some individuals, who take a real interest in the welfare of the little fish.

It is not only necessary to supply food regularly, but to see that the fish get it, and that all get their share. It will be found that some of them rapidly outstrip the others in growth, and begin ˙ to " bully " their neighbours. This makes sorting desirable, as these larger fish will, if allowed to remain with their fellows, develop cannibalistic propensities, and devour a goodly number of them. When they begin to do this they grow rapidly, and soon thin down the numbers of the other fish, so that the owner of a pond may sometimes be surprised to find on letting it off, and counting his stock at the end of the season, that there is a great deficiency, far over and above the number which have been recorded as taken out dead from time to time. These large fish, too, receive harm by being allowed to remain with the smaller ones. They naturally grow more and more cannibalistic, and would, if allowed to do so, eat up all the small fish in a pond, and then begin to eat each other. Many large trout are developed in this way, but it is doubtful whether they are desirable fish for breeding purposes.

No question can be of much greater importance to the fish breeder than the development of his stock fish. When yearlings are being sorted from time to time, as they are thinned out a few of the choicest and best-made fish should be selected and placed in a pond where they can be taken care of. A pond sixty feet long, eighteen or twenty feet wide, and four to six feet deep will do very well, and such a pond will hold fifteen hundred yearling fish, provided always that there is an ample supply of water. In selecting these fish, great care should be taken to secure those of the most perfect shape, and it will be understood by any one acquainted with cattle or poultry breeding that this is a matter that requires a considerable amount of judgment, as well as a good knowledge of the construction or build of the fish themselves. Some are at once rejected, being too long or too short, or having misshapen fins or opercles, or being otherwise not quite up to the mark. These are at once disqualified for breeding purposes. They may be perfect fish in the ordinary sense of the word and from an angler's point of view, and are perhaps even above average, but they are to be carefully kept out of the stock ponds, which are afterwards destined to produce fish in abundance for many waters.

In a state of nature trout often develop considerable deformities, and here we have the advantage, by careful selection, of avoiding much of the risk of producing monstrosities and malformations, such as are found naturally in many localities. Cases have been recorded of blind trout occurring extensively in some natural waters, as for instance in the Fischau, near Mandorf, in Germany. I have met with cases myself, both of total and partial blindness, and instances have been recorded in which trout were found to be blind in one eye only. Probably the last-mentioned cases have been due to external causes, but I am satisfied that instances of blindness occur amongst trout which are the result of disease or hereditary weakness of some kind. Blindness is well known to scientists to occur amongst other fish also.

Deformities of the jaw and head are not uncommon in some lakes, and, from experiments which I have carried out, I am satisfied that in-breeding is one of the causes. Sometimes the upper jaw is arrested in its growth, and becomes much shorter than the lower one, whilst sometimes the lower one is similarly affected. In some cases the lower jaw becomes so fixed that the fish cannot close its mouth, and occasionally instances may be met with in which the opercles or gill covers, instead of lying close, as they should do, stand out at a considerable angle, and when fish so affected are looked straight in the face, their faces have the appearance of being surrounded by a sort of Elizabethan frill. In the island of Islay is to be found a race of tailless trout, and the hunchbacked trout of Plinlimmon, in Wales, have been alluded to by many writers. These and some other deformities may be caused by the fish being carried over waterfalls, as has been suggested, but, on the other hand, there are many streams of the same kind in which such cases do not occur. It is beyond dispute that a fall from a height is liable to injure the spine of a fish, and some crooked spines may be produced in this way. This, however, would not account for the deformities of heads of fish, and for the deficient gill-covers which have been found to exist.

"Remarkable malformations are observed in the trout of

Malham Tarn, and in a beck on the western side of Penyghent. This is manifested in the former by the deficiency in the gill-cover in about one in every fifteen fish caught. . . . In the case of the ground trout of Penyghent as they are called, Mr. John Foster informs us that the malformation consists of a singular projection of the under jaw beyond the upper. These abberations are considered to be the result of inter-breeding, due to an extreme degree of isolation" (*Yorkshire Vertebrata*, p. 127). It is quite likely that deficient gill-covers are in many cases due to in-breeding; certain it is that they can be caused by an attack of fungus when the fish is young, or by an epidemic of gill fever. How far in the two latter cases the deformity would prove hereditary we have at present no means of knowing, as such fish are never kept to breed from. I have had opportunities of watching to some extent cases of head and jaw deformity, and have known them developed in fish that have passed scrutiny as yearlings, but have been deformed at the two-year-old stage.

That some deformities are not apparent until fish are well advanced in life is certain. I have tried the experiment of breeding from deformed parents, but the progeny showed no trace of the deformity. On the other hand some fish culturists have succeeded in cultivating races of deformed gold fish which are now well known in our markets. Here the deformity is hereditary, and we find the same to be the case with some birds, the different varieties of fancy pigeons being produced from the wild rock dove (*Columba livia*). Amongst domestic fowls and animals the same tendency develops, and we know what great importance attaches to it. An exhibitor at one of our poultry shows failed to obtain a prize for a very fine bird, which to him and some of his friends seemed perfect in every point. He appealed, and some discussion followed, when the reason was given for withholding the prize. The bird had a crooked breast-bone, and was entirely useless for breeding purposes. It was otherwise perfect, and this point had 'either been overlooked by the exhibitor or he had hoped it would not be noticed. Amongst cattle and horses, too, we know what importance is attached to the different "points," and what an amount of skill and judgment is required in the breeding of pedigree stock. So amongst trout, too, much care cannot be exercised in the selection of the fittest, and in their careful treatment afterwards.

The fish, during the yearling stage, are peculiarly adapted for selection. Any cases of deformity at birth, which might easily pass unnoticed in the fry stage, may now be detected by an expert, and a good selection may be made. Formerly, the largest fish were selected, but it has been found that these are often far from being the best to keep for breeding purposes. In the first place, more than an average number will probably turn out to be males, and many of them have grown big by living on their companions, and so are undesirable stock to keep, as the habit is sure to break out again. A very good way of dealing with badly-formed fish is to turn them into a pond by themselves, feed them well till they are two-years-old, and then eat them or market them for that purpose.

I have drawn special attention to this important subject for the benefit of future fish breeders. For the purpose of stocking waters the case is very different. Just in the same way that ordinary fowls in a farmyard would often be of no consideration at a show, or for breeding pedigree stock, yet do good work for their owners, so the ordinary or yearling trout, if bred from a good stock are excellent for stocking waters for angling purposes. It is enough for such purposes to lay down ova taken from good fish. In the past, any eggs from any fish have often been taken, but this should be altered if breeds of trout are to be kept up to the mark.

In sorting the yearlings, it will usually be found that in some ponds at least there are two distinct classes of fish. One class is to be found at the head of the pond where the water comes in, and the other at the lower end of the pond where the water runs out. Often the difference between these two groups of fish is considerable. Those dwelling at the head of the pond live in a good current of water, which improves even in passing along the raceway, and drives more oxygen into the upper than the fish can get at the lower end, consequently these fish have a better chance in life, and they take advantage of it by feeding better, increasing bone, flesh, and muscle, and are better made and livelier fish altogether. They are more than double the value of the others and should be carefully separated from them in sorting. Of course, some judgment is required, as in all probability some of the best fish will have scattered during the netting, and so will turn up at the wrong end of the pond. The other class of fish, if put into a pond by themselves and given plenty of good water and food, will pull up their condition and improve beyond expectation with proper care.

August is a good month for dealing with yearlings, which are now for the first time called by that name, and by the end of the month or in September some may be transplanted, provided the water for which they are intended is not too far away. It is true they are not a year old, but they have reached a very important stage in their development. They have already passed through three phases or stages of their existence, viz.:—the ovum or egg stage, the alevin or yolk sac stage, and the fry stage, during which latter they commenced to take food by means of their mouths. The last of these stages is passed in spring, and it is one during which they will travel long distances with perfect safety. After this, however, comes a period when they will not bear removal, chiefly owing to the season of the year, and this is during the months of June and July, and part of August   Not only so, but unless they have been treated with skill, a large portion of them may die during the months of May and June. To bring the fish over this stage of their existence is one great anxiety of the trout grower, and by July the survivors are safe, the dangerous time has been passed, and the fish are ready for transit again as soon as the weather will permit.

To places that are within easy distance they may be sent in September, and often even during the last days of August, and the advantage of the early planting has been already alluded to. The fish have indeed reached another stage and are yearlings, by which name they continue to be known until the weather becomes too warm for moving them in the following spring. During the same period the fish that were known as yearlings last season now become two-year-olds. ·

Yearlings are decidedly the best fish to use for stocking purposes, taking all into account. The objections to fry have already been stated. They do not hold good as regards yearlings, which have reached a stage when they can take care of themselves, hunt for their food, and feed well when they get it. They are also much more capable of avoiding their enemies, and do not run the same chance of being eaten that fry do. Next to eyed ova yearlings are, I should say, in the long run, the most economical. Two-year-olds are larger, and are in some instances available for angling sooner, but I would back a good yearling against a two-year-old in many cases. They are more easily transferred, are not so much affected by the journey, and become more readily acclimatized, as it were, to their new water and surroundings. It is true two-year-olds are larger, but the cost of transfer is very much greater, and the price of the fish is necessarily much higher to begin with. Good yearlings turned out in autumn, where there is food in the water, are almost, if not quite, as good as two-year-olds turned out in winter. The cost is certainly much less.

Yearlings require much less preparation for a journey, and therefore do not receive such a check to their growth as two-year-olds. They travel well in either metal or glass carriers, and in warm weather in spring the glass carriers have several times won the day. I have often sent yearlings to the Hebrides, Orkneys, Cornwall, and to distant parts of Ireland, and they travel as a rule without loss. Of course, with such delicate beings a mishap sometimes occurs, but it is only one in a crowd. It happens so seldom, indeed, that it is quite unlooked for, and, when it takes place, is invariably due to some very exceptional cause. Some trout, for instance, were once put close to a large fire in the baggage room, by some well-meaning porter, who "thought the poor things would like keeping warm." What the result would have been had they not been discovered and removed I need hardly describe. Occasionally, careless shunting may cause a few deaths amongst fish that have their heads to the carriers, but these are very few. When yearlings are to be sent away a fine net is run through one of the ponds, care being taken not to lift too many fish at once. It is easy to slack away the net a little before lifting, and so let out a goodly number if necessary, and lift the remainder. A thousand fish is a sufficient number to lift out at one haul, and as three or four thousand fish will often be in the net at once, this means allowing a good many to escape for the time being.

Even in an ordinary earth pond, the fish may be taken out by a few successive hauls of the net, till only about a hundred to a hundred and fifty are left. The plug may then be drawn and the water run off, and the few remaining fish are easily removed when it goes down. In a concrete pond the fish have not the same opportunity of eluding the net, and with care every fish may be lifted out. If intended for transference to a distant water, the fish should be placed in tanks for awhile, the time varying according to the length of the journey, the season of the year, and the temperature of the water. A good deal of the preparation for a journey can be done in the pond, and it is very important to have the fish just in condition for transit, and to bring them into this condition by a gradual process. Years ago, the removal of trout from one place to another was a difficult work and usually involved heavy loss. Now it is comparatively easy, and can be effected without any loss worth naming. More often than otherwise, not a single dead fish occurs during a long journey. Sometimes there are a few, and considerably under five per cent, and these cases are due probably to the shunting on the railway. From one to three days in the tanks is nowadays usually sufficient, that is if the fish have been worked into condition whilst in the pond. By having them ready counted, and the travelling tanks all ready, a large number of fish can be packed and forwarded in a very short time, which is important, inasmuch as it shortens the time occupied in transit. Orders should however always be booked well ahead, so as to allow ample time for the preparation, and the time of forwarding should in all cases, where practicable, be left for the sender to decide.

One thing purchasers should always remember, and that is, that when fish are ready to travel, that is, after their preparation, nothing should be allowed to alter the forwarding arrangements. I have occasionally received a telegram at the last moment, "Don't send the fish till next week." This may suit the sender of the message, but it certainly does not suit the fish, which must suffer seriously, and probably some of them will suffer so much that they will never recover from the effects of such treatment. They can be retanked and kept for a few days it is true, but at the end of the time will be out of condition for travelling, and

PREPARING YEARLINGS, SOLWAY FISHERY

will, therefore, suffer materially in transit. In a case of emergency, such as the purchaser being called from home suddenly, someone else should be appointed to attend to the fish on arrival at their destination, and in cases where this is impracticable an easy way out of the difficulty is for a man to accompany them who is qualified to see them properly attended to.

The net used for taking yearlings may be constructed in several ways. The ordinary yearling net is simply a seine which is made large enough to bag considerably in being drawn through the water. Mosquito netting makes a good material, and a brass chain with a few leaden weights attached is useful for the bottom edge of the net. A good heavy weight, say about four pounds, should be attached to each end of this chain; that is at the bottom corners of the net, so as to keep them down. No floats or corks are needed. Another very useful net is one attached to a large iron rim and suspended by four cords from a pole or handle. This is let down to the bottom of the pond and the fish driven on to it, or fed over it, when it is suddenly lifted, full of fish. This is the best method for ponds containing a large number of trout, as it does not disturb the bottom at all, and there is no fear of getting a lot of mud or other objectionable matter in the net, which may sometimes happen when using a seine or draw net. The contents may be discharged into a tub or some other convenient vessel placed at the pond side, and from this tub the fish may be poured into buckets, the counting being done as they pass over, and the sorting may be managed by means of a small hand net. At this stage they must be manipulated very delicately, and should be handled as little as possible. A novice will make considerable bungling at the work, and will probably injure some of them, but with practice all can be done quite easily and without doing any harm whatever.

At this stage it is not necessary to give the fish doses of earth as has been recommended for the fry. They are so much larger and stronger that they themselves stir up the matter that is at the bottom and on the sides of a pond by their movements, and in this way they get the benefit of the earth that is at hand. The bottoms of the ponds should be kept as clean as is practicable.

s

The fish will now do very much of this work by disturbing any deposit and sending it on to the screen, whence it may be raked out ; indeed, they often scour the bottom pretty well. There is a good deal of material to settle to the bottom, such as the sediment brought down by the current, dust from the atmosphere, splashings from heavy rains, the excrement from the fish themselves, etc., and care should be taken that masses of filth do not accumulate unduly. These matters are easily kept right by constant attention from the beginning, but should accumulations be carelessly allowed in the ponds, wholesale disaster may be the result.

I have sometimes been asked whether yearlings reared in ponds and then turned out can hold their own against the wild fish. Undoubtedly they can. I have tried many experiments by way of testing this, and am quite satisfied about it. Should any-one have any doubt on the subject, nothing is easier than to place a few large trout for a short time amongst any yearlings to be turned out. The latter are soon educated, for they value dear life more than might be supposed.

I turned a number of yearlings into a pond in a natural stream which flows close by my writing room, and as I now sit writing this, I can see what goes on in that pool. There was one big trout in it. The yearlings when once fairly settled took up their positions, and waited for anything in the shape of food that was brought down by the current. The big trout did the same, and I saw him several times make an attempt upon the life of one or other of the yearlings, but the way in which these eluded him was instructive. They were together in this pool for some days, but nothing serious seemed to happen till a flood took place, during which the big trout disappeared, but the yearlings still held their own, with an accession to their numbers. They might or might not be the same fish that were in the pool before, but there they were.

Yearlings are at first largely fed in the ponds by means of the feeding box described for fry, the only difference being that the perforations at the bottom of the box are larger. Some of them will, however, take sufficiently large pieces of meat by July or August to make it safe to throw the food with the hand, and as soon as this is found to be the case this mode of feeding should be adopted. It has the advantage of being less trouble. The times of feeding, as the season advances, need not be so frequent. No definite rule can be laid down as regards the number or times of the meals, but as time goes on they may be reduced to three a day, and finally to two. As the fish get older they feed better and take larger pieces of food and more of it at a time.

The daily routine goes on, and the fish become two-year-olds. They are, of course, now in larger ponds, say about sixty feet by eighteen, and four to six feet of water. If only four feet has been run over them as yearlings, it should now be increased to six feet, and they will grow all the better. The treatment of the two-year-olds is very much the same as that of yearlings, only that things are on a larger scale. The ponds are larger, the fish are larger, the water is deeper and more of it and the food is coarser, and is fed to them more freely. At this stage they may be fed partly on biscuit, which is about as economical as any food that can be provided, but care should be taken to have the right kind. I have seen some kinds that do great injury, and have therefore made special arrangements to have a biscuit manufactured that is fit for trout food, and it has more than answered expectations. At seventeen shillings per cwt. it is cheaper than horse beef, as that is the dry weight, and there is no waste whatever. It possesses also the great advantage of keeping good for any length of time, which is of importance, one of the difficulties of fish culture being the liability to run short of other food during the summer when it is most wanted, and when it must be obtained fresh almost daily. It also makes a very excellent dog biscuit, and is good for poultry or pigs, so that there never need be any loss upon it, even if not all required for fish food.

Two-year-olds may be taken from the ponds either by means of a draw net or seine, or by a dip net, and of the two methods I prefer the latter. The one I use is circular in shape and is six feet in diameter, and answers all purposes. It also does very well for lifting so-called coarse fish such as perch, roach, carp, tench, minnows, etc. Altogether, two-year-olds are a most satisfactory size of trout to keep. They need about as little attention as at any stage of their lives, feed well, and there ought to be no loss amongst them worth naming. A greedy individual will occasionally get choked, perhaps, by trying to swallow something which he should not have attempted, or one may commit suicide by jumping out of the pond. Such cases are, however, few in number, and on the whole nothing alive is perhaps much safer than two-year-old trout, when in a suitable pond with plenty of good food and water.

CHAPTER XIV.

MANAGEMENT OF MATURE TROUT.

Maturity considered—How mature trout are dealt with—The net—Its use—Emptying the pond—Business pond differs widely from a lake or river—Trout eating trout—Sorting the fish—Food—The maggot factory—Tadpole rearing—Frogs and Toads considered—Trout get very tame—Approach of spawning time—Can trout hear—Do fishes sleep—The senses of taste and smell—Varieties of colour and markings—How many species—Selection and crossing of races—Trout anadromous in New Zealand—Reversion to type—Square tail and forked tail.

THE management of mature trout is a subject on which a book might and probably will some day be written. A trout is said to be "mature" when, according to its sex, it either sheds its milt or deposits its ova. This has, I think, been the generally accepted explanation of the term. It does not, however, quite convey the idea that a fish culturist would have of maturity. I have seen yearling trout *(fairo)* yield milt, and two-year-olds sometimes yield ova. Neither of these fish are considered mature, however, on a fish farm. Three-year-old females will spawn, but the eggs are small, and not so good as those from older fish. It is in reality not until a female trout is four years old that it is of much use to the cultivator. It then yields a return in the shape of eggs which are worth having, but those from older fish are better.

We have traced the growth of the fish from the egg up to the two-year-old stage, and when it passes beyond this it has done with what is a very important part of its life history. I allude to the time during which it may be sent long distances at the proper seasons of the year without any special difficulty, a time, in fact, during which Nature seems to have made special provision for accommodating the wants of man. Up to two years of age trout can easily be tranferred from one place to another, even though the places are hundreds of miles apart. After that age the difficulties and cost of transit increase so materially that larger trout are seldom travelled. Not that there is any very real difficulty with them when in the care of an expert; it may be resolved more into one of cost. I have never met with any serious casualty in conveying large fish, but the labour and constant attention necessary are often considerable. The care of so-called large trout continues to be very much the same as they get older, and I have often compared the work of the fish-farmer with the work of the sheep-farmer. The trout are now counted, herded, sorted, marked, spawned, dipped, etc., very much the same processes that sheep are put through, and each requires its own peculiar attention.

Let us take a pond of three-year-old trout in hand in autumn. A net must first of all be run through it. Let it be a fine-meshed one, say about thirty-two meshes per foot, counting from knot to knot. It is true a much coarser net will hold the fish, and will be lighter and easier to work, but the larger the mesh the greater is the danger of some of the fish getting their heads fast. Unless it is as I have described, when one mesh gets broken the hole is big enough for a trout to get his nose in, and he will not be to blame if he does not soon get his head forced through. When a fish gills himself the best course is to cut the net away and liberate him; the net is easily mended, the fish is not, and he is sure to be injured if not speedily released. The size of the net used must of course depend upon the size of the ponds. Take care to have it deep enough. It should be at least twice the depth of the water. This causes it to bag more, and with some good weights on the bottom rope, it will work very well. A couple of extra heavy weights are required for each end of this rope, that is for the bottom corners of the net, and when hauled through the pond it should be done slowly, so as to avoid lifting these weights off the bottom. When the place for lifting is reached, it should be done as quickly as possible. With a fine-meshed net which is at least twice the depth of the water, and in length sufficient to sweep the pond, two-thirds of the fish should be enclosed at the first haul, if the work be adroitly done. It is best worked by four persons, although it is often done by two. Anyone can work the upper or cork line, and a couple of boys do very well. The bottom ropes should be managed by two men who understand their work, and at a given signal the two heavy end weights should be hauled up quickly, and the whole of the lead line got above water as speedily as possible. The cork line is then thrown well back so as to give the fish room, and they are in a huge bag from which there is no escape. They may be lifted out by means of a large landing net and placed in tubs or tanks, a net-full at a time, the remainder being left in the large net until the lot in the tub has been dealt with. When all are sorted and disposed of, the big net should be run through the pond again, and the process repeated. It may be desirable to draw it a third time; it does not take long, and is the best way of clearing the fish. A few will manage to elude it after all, but they should be very few. There will be room for a fish to get round the end probably whilst being dragged through the pond. There need not be, if properly worked. It is a good plan for a boy to walk on each bank just behind the net, and to splash the water with a stick, to deter any fish from running back. A few will probably get under the net when the lead line is being lifted. This makes it imperative that the bottom rope be raised as quickly as possible. When all is done properly there will be very few fish left in the pond.

The water may now be let off by drawing the outlet plug, and when it has run down to about a foot or less, the few remaining fish can easily be removed by hand nets, and the pond may be cleaned out and refilled. The cleaning out is a very trifling business, and is usually done by one man in about two hours. There should really be very little to clean out. A little soil must get blown in during the year, and a mischievous rat or mole may be guilty of turning some in, and there is the excrement from the fish, but otherwise there should be nothing. When the pond is to be re-stocked with yearlings great care should be taken that every fish has been got out. Should one be left lurking in a corner, or in a hole in the bank, I need hardly point out what is likely to happen before the next season's sorting time comes round.

It is hardly necessary to point out that a small pond used for business purposes, and stocked with as many yearlings as the water will maintain, differs widely from a lake or river. In the latter the fish are free to roam where they like, and soon take up their positions and look well after themselves. A few undoubtedly get eaten, but not many. But in a small pond which is crowded with fish that are well fed at least twice daily, it will be apparent that one or two large trout will reduce the numbers of the rest. The big fellows, under such circumstances, will become regular bullies, and will, to an extent, prey upon the smaller ones, which are very much at their mercy, as they have no stones to dodge behind or hiding places to get into. It is surprising, though, how the yearling fish, even under such circumstances, will avoid being caught.

Should a large trout by any means get into a pond full of yearlings, he must at once be taken out, and usually there is no difficulty in doing this. A rod and line is the remedy, and for a lure something that is too big for the yearlings. By skilful hands the big fish will soon be hooked and landed without injury.

Now to return to the fish in the tub. They must not be left there a moment longer than can be avoided. They have to be sorted, and there will be four kinds or classes, therefore four tubs or other vessels will be required in which to place them. They may be classified thus :—

I. Males.

II. Spawning females.

III. Non-spawning females.

IV. Small fish.

The *males* are now readily distinguished, and should be separated from the females, and put into a pond by themselves.

The *spawning females* should be placed in another pond, ready to be overhauled again when ripe.

The *non-spawning females* may be placed in a tank *pro tem.*, and then put back into the pond from whence they came as soon as it is again filled, unless it is desired to stock it with other fish. In any case, a pond should be in readiness to receive them, and the spawning females may be put into the same one later on, when the eggs have been taken from them.

The *small fish* are those which for some reason have not grown like the others, and are now very much less in size. There will often be a few of these, and the best way of dealing with them is to turn them out into some stream or lake, unless there happens

to be a pond available into which they can be put. Pond room
is usually too scarce on a fish-farm, but should there be room for
them anywhere they will make good fish if properly attended to.

Sometimes a net has to be hauled under the ice. This is
quite easily done. It usually happens that the water does not
freeze much at the inlet end of a pond, and as those which mean
business are rectangular in shape, the matter is a very simple one.
Cut the ice along the sides of the pond, and across the outlet end
where a slit should be made wide enough to allow for the insertion
of the net. As soon as it is set the cork line should be slipped
under the ice, and it can be hauled along quite easily, the ropes
working in the slits made along the pond sides. At the inlet end
there is usually plenty of open water to allow of the lifting of the
net, and if not it is easy to remove a slab or two, or even to break
the ice up, but not the slightest difficulty has ever arisen. The
most dangerous effect of ice on a pond at such a time, is that it
settles down when the water is let off, and without care and watchfulness a fish or two may get smothered by being left underneath
it, especially should there be any mud on the bottom of the pond.

The male fish having now been separated, should be kept to
themselves hereafter, as they do very much better away from the
females. They do not fight so much as they do when mixed
together, and are not so apt to be attacked by that well-known
scourge, fungus *(Saprolegnia)*. The food, too, can be varied a
little. It is a great mistake to suppose that the males do not
require well feeding. At the same time they need not be allowed
to share all the rich and expensive food which it is necessary to
supply to the females. The question of food for the large fish is
a serious item, and is one of the heaviest expenses attached to the
growing of trout. In addition to the beef, biscuit, mussels, etc.,
a large quantity of natural food can be produced on the farm. All
the scraps of meat, sinew, gristle, and bone, may be largely made
use of; indeed, nothing should be wasted. A great deal of this,
as well as the numerous rats and other vermin which should be
trapped, may be converted into maggots, which are excellent diet
for the fish.

There are many clumsy ways of making these, but they are
of little use practically. One or two proper maggot factories

should be built.   They are simply wooden houses, with an upper
compartment, into which the meat should be put.   The bottom
of this chamber ought to be made of wooden slats, or, if preferred,
an iron grating can be used.   The object of this will be seen at
a glance.   It is for the maggots to drop through.   A wooden tray
is placed underneath to receive them, and this tray should be
about two or three inches deep, and two of its sides should slant
outwards at an angle of about forty-five degrees.   In addition to
maggots it receives also the juices which drop from the meat, and
it should, therefore, be made perfectly tight.   The maggots crawl
easily over the sloping sides, and drop down on to the bottom
floor of the maggot house or factory, which should be covered
with an inch or more of fine dry sand.   The grubs roll themselves
in this, and become thoroughly cleaned, and are raked out into a
receiver of any convenient kind, and the sand riddled out, dried, and
used again.   Then they are put into a tub or a box and mixed
with a fair quantity of fine meal and dry sand.   In the course of
twenty-four hours the meal is consumed by the maggots, which
are very much improved thereby, and the dry sand with which it
is mixed prevents them working it up into a paste, and also helps
to scour them still further.   The sand is then riddled out again,
and the grubs are scalded, and after this preparation look
delicious, and are most tempting morsels for hungry trout.   The
trout think so too, evidently, by the way in which they take them
when thrown into the ponds.

In addition to the maggots, which may easily be produced in
large masses, a great quantity of food can be provided in the
shape of tadpoles.   In spring many tons of frog spawn are
collected and placed to hatch in ponds about two or three feet
deep.   When the hatching takes place these ponds are black with
tadpoles, and as there are a great many more than nature can
provide for they must be fed.   The scraps of meat which later in the
season go to the maggot factory are now used for fattening tadpoles.
They are voracious little beings, and will eat any carrion which is
thrown to them.   They do not like much running water, but a
very gentle flow should be kept going through the ponds, in order
to prevent contamination by the decomposition of the meat.
They eat ravenously and grow amazingly, and in a few weeks get

as fat as seals. A short time before they turn into frogs tadpole traps should be set, several in each pond. These consist merely of a bone, each suspended in the water, or a piece of meat may be used. These baits are found night and morning to be one mass of tadpoles, which swarm around them in great numbers, and rather resemble swarms of bees. They are lifted out twice daily in a very simple manner. Take an ordinary riddle or sieve of such fineness that they cannot slip through its meshes, and tie this firmly on to the end of a short stick. An old hay rake handle is a very good thing, as it is usually split into two prongs at the end, and is more easily fixed firmly than an ordinary stick. Armed with this apparatus the man in charge goes quietly to the pond, adroitly dips the sieve under the pendant bunch of tadpoles, and lifts. They immediately leave the bait and fall into the sieve, which is then withdrawn, and the forsaken bait being tied, is dragged over the side and falls into the water again, the trap thus resetting itself. The contents of the sieve are then emptied into a pail, and another bunch is lifted in the same manner. A very large number are easily taken in this way twice daily, ánd the trout soon make short work of them when they are thrown into the ponds. This food possesses the great advantage of being alive and of living in the water, and if any are not immediately devoured they remain in the trout pond and are eaten afterwards at leisure, and they do not foul the water. Should a quantity of them remain unused until they turn into frogs, they do equally well for large trout which eat them freely. The only difficulty is that when they become frogs they will get out of the pond, and many of them will be lost unless they are sharply looked after.

Toads spawn later in the season than frogs, and the spawn consists of long ribbon-like masses, the eggs being in a gelatinous-looking mass like frog spawn. Trout will eat them, but they do not relish them as they do frog tadpoles. Immense quantities of worms can often be obtained very easily by sending children to follow the plough and gather them—indeed it is sometimes worth while to drive a plough simply for the sake of the worms. This is one of the many ways in which agriculture and aquaculture work together. Willows may be profitably grown on the waste ground, where such exists, about fish ponds, and a better use for these cannot be found than making them into eel baskets, which are profitable to work, in addition to being exceedingly useful in keeping down the trout's worst enemies. The pond mud which is shovelled out at spawning time when the ponds are cleaned is a capital manure for many crops, and amongst other things I may mention a couple of thousand black-currant bushes, which I find it suits admirably.

Trout cannot see their food at a great distance, but when a handful is thrown into a pond a few fish are sure to be near, and they go for it; other fish seeing them, get the signal, and soon the water is in a state of great commotion. They soon see the attendant who feeds them when he makes his appearance on the bank of a pond, and they get to know him. When very hungry, however, they will often come to a stranger. Colours evidently make an impression upon them. I have seen a whole pond-full of fish not only refuse to feed, but keep out of sight when a lady in a light-coloured dress or with a parasol has appeared on the bank.

I have had trout so tame that they would come and take food out of my hand, but they are rather rough in their manners, and their sharp teeth will sometimes lacerate the hand severely. Anyone trying this experiment and being successful, is not likely to keep it up long. Trout have been credited with very rapid digestive powers. I am not so sure that their digestion is always so rapid, however, and care should be taken not to overfeed them. They are certainly more liable to disease when too well fed. A little practice will soon enable the caretaker to know when sufficient food has been given.

It sometimes happens that trout will not take food. This is often notably the case during times of hard frost when there is ice on the water, At spawning time also they eat very little, and after being handled, as for instance for sorting or counting. This should never be done except at spawning time, when it is desirable to examine every fish on the farm that is over two years of age. This work should be carried out just before the spawning work commences, and should be finished before the actual "stripping" of the fish is begun. Trout eat twice as much in summer as they do in winter, and often more than this. As a general rule it may be held that with a low temperature they eat but little, and with a high one they eat greedily. There are many other influences at work, however, which cause them to eat more or less freely, but which can hardly be discussed here. It is well known that during the summer months they accumulate a great amount of fat, and thus they are better enabled to bear the starving which they get during the winter.

There has been a great deal of controversy as to whether trout can hear. The ear of a trout may be briefly described as an internal organ that is tolerably well developed, and is situated within the cranial cavity. But it has no tympanum, nor yet any tympanic chamber. That a trout cannot hear in the ordinary way is, I think, beyond any doubt, as the ear is not constructed for that mode of hearing. But trout are exceedingly sensitive to vibration, and any sound made beneath the water, or even in the earth, as for instance in blasting a rock, may be transmitted by vibration, and the fish in this case would be immediately sensible that a disturbance had taken place.

The labyrinth of the ear is well developed, but the external orifice is closed, and we know what is the effect on ourselves if we, for instance, put our fingers in our ears when anyone is talking. I once tried some experiments with fish with regard to their sense of hearing, and I came to the conclusion that trout could not hear sounds made in the air, provided no agitation of the water occurred. The same applies to some other fishes. Amongst others I operated upon a shoal of herrings (*Clupea harengus*) in a tank. Moving my hand to and fro close to the glass, I got the fish so accustomed to the sight that they took no notice of it whatever. But the moment my hand smacked against the glass they scattered in bewilderment. I tried shouting and making noises, but they apparently did not hear at all. Now, there is in many fishes a connection between the internal ear and the air bladder. In the herring, for instance, we find at the fore-end of the bladder two very narrow tubes or ducts, which connect or at least come into exceedingly close contact with the organ of hearing, passing right through the skull, and apparently connecting with the labyrinth of the ear.

There is also in the herring a very curious connection between the stomach and the air bladder, so that on inserting a tube and inflating the stomach, the air passes on and distends the air bladder also, and, what is more remarkable still, the posterior end of the bladder is found to narrow into a tube, which ends in an external aperture near the anal fin.   So that the ear of the herring may be said to be connected with an external aperture.   Many other fishes have the connection between the air bladder and the organs of hearing ; but the external aperture connecting with the air bladder seems to be absent, except in the herring.

It has been suggested that the air bladder may act as a drum, and that sounds may be transmitted from it to the internal ear. Be this as it may, the question is one of great interest, and is well worthy of investigation.   It is quite likely that many fishes which have not yet been thoroughly examined may be found to possess the same external aperture that occurs in the herring, but I cannot find any trace of it among the *Salmonidæ*.   As far as regards the members of this family with which I am acquainted (and I know a good many of them) I should certainly say that they were very " hard of hearing " indeed.   I am alluding now to sounds made in the air and not in the water.   I have often had the opportunity of testing the matter, as for instance, in the case of firing a gun close to the fish, and in many other ways.   If they see nothing they take not the slightest notice.

Livingstone Stone, the great American authority, is also very much of the same opinion, and says in his *Domesticated Trout* (p. 221) :—" I will not say that trout cannot hear ; but this I will say with the greatest positiveness, for I have tested it repeatedly, that they are not frightened at noises, however loud, nor do they pay the slightest attention to them.   You may place your mouth directly over the trout in a pond, and, if they do not see you, you may scream with all your might, or ring a bell as loud as you please, and the trout will not move a fin to show that they are either frightened or attracted, or that they have in any way noticed it."

Seth Green, in his *Trout Culture* (p. 58), says that trout cannot hear, and that " they will not stir a fraction of an inch at the sound of a gun fired one foot above their heads."   This is, of course, provided they see no flash or smoke.   With regard to the ringing of a bell, we have heard of fish being summoned to a meal by the performance, and I have often heard this story brought up as a direct proof of fish being able to hear. I once had a visitor who was so sure that the fish, when trained to do so, would obey the summons, that I at length told him that I had some ponds full of fish that would do the same, but when I explained that it made not the slightest difference to them whether the tongue of the bell happened to be in or out, even he began to be less credulous. Having procured a bell, and removed the tongue, I went out for his satisfaction and pretended to ring it, when the fish came at once.

Although no notice is taken of sounds made in the air, yet we find that trout are keenly sensible to vibrations of the water, which are probably transmitted to the body of the fish, and there is little doubt that noises made in the water are heard by the trout in this way. Dr. Francis Day, in his *British and Irish Salmonidæ* (p. 19), says:—" Hearing is developed in fishes, and it is very remarkable how any diversity of opinion can exist as to their possessing this sense." A little further on he says:—" But the chief mode in which hearing is carried on must be due to the surface of the fish being affected by vibrations of the water, and the sounds are transmitted directly to the internal ear, or else by means of the air bladder acting as a sounding board." Here we must leave this interesting question. Its bearing on the cultivation of trout is rather an important one, as cases may often arise in which the fish might be so disturbed by noises as to be injuriously affected thereby. But it is not so, for as regards noises made in the air the fish are neither alarmed when feeding, nor yet driven off their spawning beds, provided always that they see nothing.

Some scientists have supposed the lateral line to have some connection with the hearing powers of fish, hence it has been called " the lateral sensory apparatus." It consists of a series of punctures forming tubes from which nerves run, and com- municate with the head and possibly with the organs of hearing. These tubes also discharge a mucous secretion.

Do fishes sleep? is a question I am sometimes asked. That they do there can be little doubt, but whether sleep is to them exactly what it is to the mammalia is another thing. Trout cannot close their eyes, but that they rest their bodies is very certain. We know that they feel fatigue, and that they spend part of their time in holes or under shady banks. They love shade, in fact, and cannot do well without it. At night, when viewed by the light of a lantern, a whole pond full may be watched at times without their seeming to be in any way inconvenienced by the light. They are at such a time often found to be in a perfectly quiescent state, and also in the day-time I have seen a pond-full apparently sleeping. The great difference between their sleep and that of birds or animals is that they are apparently able to see at the same time. We know well enough that all parts of the body do not necessarily rest at once, and we also know that the sleep of a city merchant and of a savage are two very different things. The latter will almost sleep with his eyes open, and we can well understand a fish being on the alert, even though asleep in the sense of the body resting.

The sense of taste and smell are to an extent existent in trout. The nostrils are well developed, but have no communication with the mouth, nor are they used in any way for purposes of breathing. Probably under some circumstances the power of smell is useful to the fish in obtaining food, as for instance, in blind examples which I have met with that have otherwise been in excellent condition. The odour of water is naturally extremely varied, according to the different ingredients or pollutions which may exist in it. That trout are sensible of these variations there · is no room for doubt. They themselves give off odours and so do other fishes, and I have long had reason for supposing that eels can smell trout and follow them by scent. An otter can certainly do so as I have found to my cost, and these animals on a stream must become sensible of the presence of an unusual number of fine trout kept in ponds near, the water from which is discharged into the stream.

Trout possess in a degree the sense of taste, but it cannot be said to be at all well developed. They will accept some foods while they reject others equally good or even better. This shows that there must be a distinguishing power between one article of food and another. Fish culturists know that when accustomed to certain kinds of food they often refuse an article with which they have previously been unacquainted, but on being compelled by hunger to eat it afterwards become ravenously fond of it.

In cases where trout have absolutely refused some (to them) new kind of food, and it has not been expedient to break them on to it by hunger, I have done it by mixing a little and increasing by degrees, and then diminishing the quantity of the old food, and they have thus been beguiled into eating the new. It is quite evident that they have their likes and dislikes, as is shown by their rejection of one bait whilst they eagerly take another. Domesticated trout are fond of frog tadpoles, but dislike those of toads. There is a fœtid odour about toads which does not exist on frogs, and the trout can at once tell the difference. It is quite probable that the sense of smell has much to do with this as well as taste.

The great variety of colour and marking which exists amongst trout must have struck everyone who has had much to do with them. There are many influences at work, each of which produces its own variety. Then, again, the blending of these influences, and their consequent colouring, the crossing of different races, hybridism, age, sex, good or bad breeding, migration, season, food, surroundings, bottom, and last but not least the quality as well as quantity of the water itself, to say nothing of exposure to light or otherwise. All these and many other influences are at work, making up an innumerable number of varieties, many of which have been regarded by the older naturalists as distinct species.

Trout may be briefly divided into two kinds, viz., the anadromous, or sea-going, and the fresh-water forms. Both are more or less migratory, and therefore the terms "migratory" and "non-migratory," which have been so often used, are not altogether correct. Even some of the so-called fresh-water kinds are known to go to sea occasionally.

The question as to how many species of trout exist in Britain is one that has evidently puzzled the ichthyological writers of the past. The tendency has been to increase the number of species as new types or races were discovered, and very naturally so, considering the knowledge that was then possessed. But fish culture has put a new phase on the subject altogether, and we are

T

now finding out many things about the genus *Salmo* that were unknown and unthought of a few short years ago.

It is well known that trout in their natural haunts assume endless varieties; indeed it would be difficult in most cases to find two exactly alike. We find in a great many instances races of yellow, silvery and dark-coloured trout (often called black trout) inhabiting the same lake. In some streams which are frequented by marine or anadromous forms we often find a considerable variety, and every connecting link may be obtained at times between sea-going and fresh-water forms, as for instance, *Salmo trutta* and *Salmo fario*. In certain cases undoubtedly these fish have crossed with each other, and it may be their descendants have crossed again, thus producing considerable variations, quite enough to puzzle the older naturalists who were not fish culturists, and therefore had not the same facilities for ascertaining the real truth. When we come to deal with these fish in a practical manner, within the inclosed boundaries of a fish farm, we find out that there is in reality very little, if any, difference between the races, beyond the inherited greater migratory instinct of the sea-going kinds. They spawn here at the same time and in the same places, and I have more than once found *trutta* paired with *fario*. The mingling of the two races in our streams is inevitable, and knowing what we do now as to the fertility of the offspring there can be no doubt whatever that the two forms do at times get considerably mixed.

Occasionally, owing perhaps to some exceptional circumstance, such as rain after dry weather, or it may be a sudden change of temperature, the streams frequented by the spawning fish become crowded, and the involuntary mingling of the two races must, one would think, occasionally take place. I have seen both forms spawning in such close proximity that I am fully convinced that milt and ova of the two races must occasionally come in contact, and impregnation take place. Anyhow, we have the intermediate varieties occurring in Nature and we can bring about exactly the same results by artificial means.

We know that in the human family some individuals have a much greater propensity for roaming than others, and some races of men exist who make periodical migrations. Why should not the same tendency be found, in individuals, amongst fishes ? We know now that trout of the *fario* type will become anadromous in their habits, and we know that it is not only possible to retain sea-going varieties in fresh water, but that they will also breed there. The typical colouring of the sea-going kinds is of a silvery hue, with more or less of a steel blue or greenish cast, but when these fish enter the fresh water their colours soon lose their brilliancy, and in course of time the fish will assume, *pro tem.,* something more of a fresh-water appearance. This is notably so in the case of *Salmo trutta.*

I have trout *(fario)* in the same pond that they have occupied for years, some of which are silvery, some yellow, some brown, and some steel blue, and yet all these fish retain their original characteristics, notwithstanding the fact of their being in the same water, and having the same surroundings and the same food. One thing I have noticed, and that is that after heavy rain the water assumes a darker tint, and the fish become darker too, but they do not lose their original colours, and when the water becomes lighter again the fish become lighter also, but their general characters are not altered. A yellow trout remains a yellow trout, and silvery fish remain silvery, and well-known individuals have remained so for years. But as regards their offspring there may be, and probably is, a slight difference in colour or in general appearance, and, as this is perpetuated, a very interesting question arises as to how long it takes to establish a distinct variety or race of trout by natural means.

We know what can be done in artificial culture by careful selection, in the crossing of two good races. The result is a good breed or race of fish, combining the colours of the parents. On the contrary, if we cross two badly-developed fish, or even go on breeding " in and in " from the same stock, we are sure to develop inferior races.

It will be apparent at once that there is no end to the varieties of trout which may eventually be produced, and probably ere long the ingenuity of some fish culturists will be exercised to introduce new forms to the fancier, very much as animals, fowls, pigeons, etc., have been similarly treated.

There is one remarkable feature regarding colour that is

worthy of note, and that is that trout vary very much more in fresh water than they do in the salt. Some colours are also much more permanent than others. The experiment is now pretty well known of putting fish into a tank that is black and kept almost dark. When taken out and placed in the sunlight in a white bowl they look very black, but in a few minutes will often become light in colour. This is also largely the case with wild fish passing from lake to stream, or *vice versa*, or from one geological formation to another. Much more so is it the case when the migration is from river to sea, or sea to river; not only do the colours and general appearance of the fish become changed, but we find, in the case of fresh-water forms becoming anadromous, and *vice versa*, that the various organs and parts become altered also, probably by the law of adaptation.

It is now an ascertained fact that trout (*Salmo fario*) can become anadromous in their habits. This seems to be largely the case in some of the rivers of New Zealand, and also in those of Tasmania. In British waters instances of brook trout migrating to the sea are common enough. I have examined many brooks at low water, and have found beautiful little trout under the stones between tide marks. Quite possibly this may have an important bearing on the fine quality of the fish occurring in some lakes in close proximity to the sea, as, for instance, Loch Stennis, in Orkney.

An interesting experiment was accidentally tried a few years ago by the owner of a fish pond near the sea. He built a rather elaborate outlet of heavy masonry, and carried it forward some distance into an estuary, so that the tide rose and fell within the walls. In it was placed a grating, which prevented fish passing in and out. Now it so happened that a few trout passed down out of the fresh-water pond above, and got into the tidal tank, which always had some two or three feet of water in it, and at high water a considerable depth. The water that entered from the sea usually swarmed with small crustaceans, etc., and when the trout were discovered some months afterwards they had grown to a much larger size than any in the large pond above, notwithstanding the confined area of the place in which they had been living. This opens up a very large question as to the future possibilities of growing trout in such situations that they have access to salt water.

In most trout there exists a tendency to diverge from the original type, and yet on the other hand we find in very marked and apparently distinct races, a considerable disposition to revert again to the original type. This reversion to type is very marked in the case of some birds and animals which have already been referred to. In the trout it is so marked that the question as to how many species we actually possess comes in and claims our serious attention. We find that fresh-water forms become anadromous, and also the reverse, viz., that sea-going forms become land-locked, and that the most marked races exhibit strong symptoms of reversion to original type, and all these facts lead us to one point, and that is a common ancestry. Here we must leave this deeply-interesting question. Future investigations will no doubt throw a good deal of light on so important a subject. In a few more years we shall know how trout behave themselves in other parts of the world, as for instance, in South Africa and the island of Ceylon. Already a commencement has been made in both these latitudes, and we are hoping soon to be able to report some results. Trout are introduced, and the future will possibly develop some new and important facts concerning them.

We know that, by judicious selection and inter-breeding, races are improved and desirable varieties perpetuated, and we know also how easy it is to lose the thread, as it were, of a pedigree race by a little careless manipulation of the breeding stock. There is endless scope for selection and improvement, the making of cross-breeds, and of hybrids, and investigating the life histories of some of the sterile forms that are occasionally met with in Nature, and which are also easily produced by artificial means.

Trout do not necessarily deposit ova every year. As a rule they do, but a few do not; that is to say, a trout occasionally misses a year. There does not seem to be any hard and fast rule by any means. Probably trout ought naturally to spawn every year, but a fish which has been sickly or ill-fed, or that has got injured in some way or other, proves an exception to the rule.

Broken fins unite again, and those that have been lost are sometimes reproduced, but they always show traces of having been badly used. The fins change with age, and it is, therefore, possible that a fish may at different times of its life have a forked tail or a square tail.

A tail sometimes appears to be forked because it is not fully expanded, and the accompanying diagram (Fig. 46) will show how a forked tail may sometimes be made into a square one, by expanding a little.

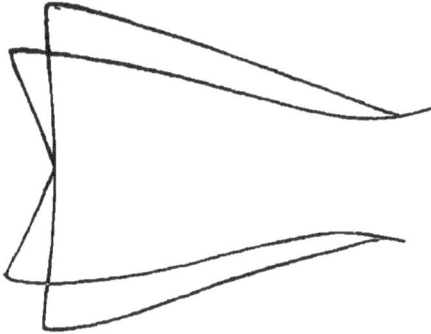

Fig. 46.

Many of the older writers have defined species by the numbers of the pyloric cæca, but these are found on examination to be ever-varying quantities, even in forms that have been supposed to constitute distinct species. From twenty-five to ninety the number of the former seems to range, the average being between forty and fifty. The usual temperature of a trout is nearly about the same as that of the water in which it lives. Red spots may sometimes be seen on trout after death that were not visible during life.

CHAPTER XV.

SALMON CULTURE.

Great loss in nature—Large number of eggs deposited—Bad management of our rivers—Some evils may be remedied—Poachers considered—Impounding Salmon —Where to get the best eggs—Nature's discrepancies provided for—More about poachers—Fate of the eggs—Falling off in catch of salmon—Rate of natural increase considered—Feeding of salmon—Migration—Experiments—Smolts and grilse—The United States—Salmon of Alaska—Alaskan and British salmon compared—Saprolegnia.

THE so-called artificial cultivation of the salmon *(S. salar)* is in a degree somewhat similar to the cultivation of the trout *(S. fario)*. That is to say, the treatment of the eggs is very much the same, as well as of the alevins and young fish. The difference lies chiefly in dealing with the adult salmon as well as the younger fish after they have been turned into the rivers, and last, but by no means least, in the manipulation of the rivers themselves. The subject is a very large one, and I can only somewhat briefly allude to it here, but as the presence of the king of fishes is required in order to make what many would consider a perfect "angler's paradise," I will give a few of my views on this important question.

If we take into consideration the fact that when left to Nature not one egg in a hundred produces a fish that will reach four ounces in weight, and make a careful examination, in order to ascertain the cause and whereabouts of this great destruction during the earlier stages, we find that enemies are so numerous and the casualties and dangers to which they are exposed so great that the wonder rather is that any come to maturity at all.

In the first place we know that salmon and trout will eat not only their own eggs, but will, if they can, eat each other; hence the enemies which the young fish have most to dread may perhaps be their own parents. It very often happens that a salmon has no sooner got rid of its ova than some hungry trout which has been lurking in the vicinity, and probably anxiously watching the whole operation, goes in and makes a meal of that which with proper care might produce hundreds, nay perhaps thousands, of fish. Thus the eggs which a salmon may travel over a hundred miles to deposit, threading rapids, wriggling over shallows, leaping weirs and cascades, and encountering all manner of difficulties and dangers by the way, are often destroyed as soon as they are shed by the parent fish.

It is estimated that at least seventy-five per cent. of the eggs are lost immediately, and of the other quantity a large portion is destroyed in one way or other before hatching takes place. We know that whole spawning beds are often washed away, or buried many feet deep by the *débris* brought down by floods, to say nothing of all the host of enemies (animal, vegetable, and mineral) that are ranged in battle array against poor unfortunate *Salmo salar*. When we consider for a moment that a single salmon deposits in one season, say ten or fifteen thousand eggs, according to its weight, and that owing to the great destruction, natural and otherwise, not one egg in five hundred produces a mature fish, and remember how at this comparatively small rate of increase some of our rivers formerly held a goodly number of fish, surely it will at once be apparent that if one-fourth of the ova annually deposited came to maturity, the rivers would in a few years be so full of fish, that it would require some extra means of capture for adequately dealing with them, and keeping their increasing numbers within bounds. And to the skilled fish culturist there is nothing unreasonable in the assumption that such a result might be brought about by the simple application of the proper means for doing it.

The work of collecting the ova, being so similar to that detailed for trout, needs no further description. That which seems to have been rather badly managed is the collection of the spawners themselves, before they are ripe, and their retention in suitable ponds. The cost of collecting ova in the ordinary way, by netting a stream, is very considerable, and, as we have seen during the past few years, is somewhat uncertain in its results. There is no need for this state of things whatever, and it is exceedingly trying to a fish culturist who is deeply interested in the work, to look on year after year and see the meagre results that are accruing to salmon-cultural operations. I have had some considerable experience in the collection of spawning fish and of ova in years gone by, and my opinion of the method usually adopted, of trying to net the ripe fish out of the rivers, is that it is most unbusiness-like. I have often been unable to get a ripe fish, and have spent days by the river-side, with men, and nets, and boats, etc., trying in vain, although catching plenty that were not ripe. I have pointed out the proper course, and the only one that is really workable, but the excuse usually is that it would be too costly.

It is the ordinary plan that is too costly. So costly is it, indeed, that for seven years or more I have not put a net into the water to catch a salmon. The remedy is to impound the fish, and when that is properly done, and the rest not left undone, then I venture to say that salmon culture will yield its fruit.

The expense is very much less than is incurred by the present system. The real difficulty is that some outlay is required before the work can be commenced, and it seems to be the fear of this that prevents it going on. Now, I may as well say at the outset that our salmon rivers are not to be put into the condition that so many people desire without a reasonable outlay of capital upon them. Without that it would be just as hopeless to attempt any improvement as it would be to attempt to put into order a piece of cultivated land which had been allowed to run waste, or, it might be, a piece that had never been cultivated at all.

The impounding of fish cannot be done by anybody. I believe it has been tried in some instances and found to fail. Nothing is more likely. The ordinary water bailiff probably knows as much about impounding salmon as he knows about harpooning a whale. Even experts have yet a great deal to learn about this, as about many other things, but that is no reason why important work should be left undone. That salmon can be impounded, and impounded successfully, I know from my own experience, for I have tried a good many experiments with the king of fishes, and would have tried very many more but for the way in which he is hedged about, in such a manner that some of his best friends are unable to help him.

We have done so much to injure both the salmon and his hunting ground that it is surely time we did something to make amends for past grievances, and to show our appreciation of the gallant efforts which he has made to hold his own in many of our rivers. We have emptied the rivers during many months of the year by draining the hills; we have rendered the water distasteful to him by polluting it. We have altered the temperature of the estuary by changing its nature; we have left him in pools to be destroyed by poachers, instead of giving him a friendly lift up or down, as he may require. We have done many other untoward acts, and it is about time that we sought to remedy some, at least, of the evils of the past.

The water question is by no means such a difficult one as may at first sight appear. The advantages which may be made to accrue by letting off compensation water from reservoirs made on some of the upper tributaries of a river are too great to pass lightly by. This question of water storage is indeed one of the greatest importance, and entitled to the fullest consideration. The immense benefit arising from it does not need pointing out. To be able to make an artificial spate just at the right time is a power which might be exercised on a salmon river most beneficially; indeed the effect is at present almost beyond calculation.

On one river which I was called in to inspect, I saw a pool from which sixty salmon had been stolen in one night but a short time before. An artificial spate of a few hours' duration would have helped these fish over the barrier which obstructed them. The loss to the river it is difficult to estimate. The intrinsic value of the sixty fish might perhaps be represented by as many pounds sterling, which alone would have paid a watcher a year's wages; but when we consider that these were spawning fish, and that their eggs and future offspring were lost to that stream absolutely, we begin to realise, after some reflection, that the loss was very serious. It might easily have been prevented. This is but one out of many cases that have come under my notice in one way or other.

On another river I was shown a pool in which many fish were frequently imprisoned, and where the owner did not wish to incur the expense of a fish-way. These fish were also taken out by poachers. Nothing could have been much simpler than the best method of dealing with this pool without making a pass. The fish were very easily taken out, and I advised that they should be lifted over the obstruction. A horse and cart a few times a year would have done the work at a very trifling cost, and would have saved the fish, or if preferred the stream itself could have been made, by means of a wheel and elevator, to lift over every fish as soon as it came up. I am satisfied of one thing, and that is that it would often pay a great deal better to lift the fish over an obstruction than to watch them, for the salmon poacher is up to all sorts of tricks, and often pays more attention to watching the movements of the bailiff than of the fish. He can quickly deal with the latter when he gets his opportunity, and it is for that he watches. It is all very well to say that poaching is kept down on this or that river, or that the poachers do not get many fish. I know the case is often very different, for I have made a special study of the poacher, his habits and handiwork, and he is not by any means to be despised.

One great advantage that is to be gained by impounding the fish is the opportunity it gives for the selection of the fittest. Often under the present system eggs are so difficult to procure that any fish that come to hand are gladly taken, and it sometimes happens that milters cannot be obtained, and eggs are lost or go unimpregnated, to give endless trouble in the hatcheries afterwards. One man once told me very seriously that he had under such circumstances used the milt of a trout, and he evidently thought he was giving me a good wrinkle by tendering the information. I heard of another somewhat similar case in which sea trout milt was used. The sooner the work is put upon a proper basis the better will it be.

I have often known salmon ova to be collected from any part of a stream where the fish could be got at most readily, sometimes being taken from the extreme head waters, and sometimes close above the tideway, or even in the brackish water itself. Salmon will occasionally spawn in brackish water, for I have seen them do it, but the eggs will in such a case probably be lost to the river. And it is better that they should be, for they would most likely only have produced a poor lot of fish. There is such a strong migratory instinct implanted in the salmon, such a desire naturally to push up rivers, that when I see them just entering a river and then depositing their ova, I am led to the conclusion that probably some of these may not be the best fish from which to propagate. I have observed something of the kind amongst trout, fish that are not in "good form," and will in consequence deposit their ova anywhere, so to speak.

Then as regards the fish that push up to the extreme head waters of a river. The migratory instinct may be a good thing, but there is an old saying that reminds us that it is possible to have "too much" even "of a good thing." My experience is that in some rivers at least these fish are late spawners. They are often also small in size; it is perhaps somewhat natural that they should be so. Their size and the extra energy which they exhibit points to the fact that they may be young fish, and fish culture teaches us that some young fish spawn later than the older ones. We know also that eggs taken from young fish are not so good by a great deal as those taken from middle-aged fish. Such points as these seem to have been very much overlooked in the past by many collectors, and yet to the fish culturist they are known to be of vital importance.

Times and seasons for collecting ova have often been beyond consideration ; that is to say, because eggs could not be got at the right time and place, owing perhaps to floods or other causes, therefore the hatching boxes have been filled with any eggs that could be obtained, irrespective of the source from which they came, or the time when they were taken. Now if there be any lesson that fish culturists have learned it is that this sort of thing will not do. I have learned it for one, and that years ago, and at considerable cost to myself, for I have had the battle to fight very often single-handed, whilst those who ought to have rendered help often ran away instead.

To get on the "wrong tack" in fish culture often means to get far out of one's "course"; indeed, there are many who have held on until they have found themselves on a "lee shore," and this is just why I wish to give a note of warning to those about to ·enter upon the work. " Do nothing rashly " is a motto that it would be well to observe in time. Nature provides for everything ; not one little bit of mechanism is wanting, and we find the result perfect. Take the flowers of the field or the fish of the sea as examples. Any one of them is perfect in itself. It remains so, and from generation to generation there is no perceptible difference in a species, be it what it may. But Nature makes special provision for this being so, and, unless we do the same, confusion will be the inevitable result.

It is true that we find monstrosities and malformations in the natural world ; it is according to the laws by which it is governed that it should be so. But we only find them perpetuated under special and peculiar circumstances—they may be called isolated cases. Take, for instance, the tailless trout of Islay, or the hunchbacks of Plinlimmon, to both of which reference has already been made. These are cases of fish living in pieces of enclosed water where they are very much separated from the rest of their species, and this probably accounts for the apparent discrepancies, if they be rightly so called, but they end there—they can get no further. We find, then, that Nature provides for such discrepancies, by arranging barriers beyond which they must not pass ; and, if we take the trouble to investigate, we find that provision is made for preventing them from spreading, and for the preservation of a perfect race or type of fish. Nature, indeed, goes in for the selection of the fittest.

This, then, is what we ought to aim at in our work of cultivating fish of any kind, viz., the preservation of a perfect race or type. In order to do it, it is clear that we must take some pains to secure the best breeders. Having advanced so far, we have the power given to us of still further improving the stock by judicious crossing of the fish of different rivers. There is a large field open for investigation here. We know the great advantage of having good breeders amongst trout, and also of introducing new blood, and there is no reason for supposing that salmon should be made any exception to the rule which applies in such a marked degree to trout.

Let us turn our attention again for a few moments to those

fish that push up to the very sources of some of our rivers.   Let
us follow them.   First, they are blocked in some pool half-way up
stream by an obstruction that is impassable except during a
heavy flood.   What takes place?   Why, half of them are taken
out by poachers, and maybe even a larger proportion.   The
wholesale or professional poacher, call him what you will, attends
to his work.

Then after a while a friendly flood comes down and helps
the remainder of the fish over the barrier, and they are soon
scattered over the head waters of the stream.   The flood subsides,
and the water is soon in the other extremity.   The rapid torrent
becomes a succession of clear pools with fish in them, and so
little water that they are unable to pass from one pool to another.
Another class of poacher now appears, and he is the individual
who would not go far out of his way to catch a salmon, but who,
on seeing one under a bank or in a pool, and a pitch-fork handy,
considers that he has as much right to the fish as anyone else, and
carries it home under his coat.   So that some of these fish never
reach the spawning beds at all.   But suppose a fourth of them
get there.   They have escaped their greatest enemy, but they
have not forgotten him.   They do things in a hurry, as I have
found on many occasions, and the eggs are often washed away, or
buried, or left dry.

But after all some of them reach the hatching point—What
then?   They hatch in spring, and, having absorbed their sacs, if
spared to live so long, they start upon a journey.   I have watched
them coming down the streams, and have seen how in every pool
they have to run the gauntlet.   Hungry trout waiting their prey
get fully half of them, the birds pick up a few; all the way down
for miles they run the risk of being devoured by eels and other
fish, and probably very few indeed ever live to return to the river
as mature salmon.   Eeels should be well looked after on every
salmon river.   They are well attended to in New Zealand, I see,
by one of the reports lately to hand.

There is a serious falling off in the catch of salmon on many
of our rivers, and when this is the case there is usually a cry made
for the extension of the fishing season, which, if allowed, would
only tend to reduce the stock of salmon still more.   On the other

hand, suggestions are made in certain quarters that the fishing season be curtailed. The way of escape is by cultivation. This is the key to the situation. The great fecundity of the fish, and the enormous loss of life that takes place if left to Nature, at once point to this, and now that we know how to deal with the matter there is good reason for engaging in the work.

The guillemot and some other sea birds lay but one egg, or rear but one young one in a season, and yet they are extremely abundant in their localities. The wood pigeon lays but two eggs, and probably a pair of birds will not succeed in rearing, on an average, more than two or three couples of young ones in a year, whilst an elephant is said to have but half-a-dozen in a lifetime of a century. Now, the young of animals and birds are cared for by the parents until they are really capable of caring for themselves. A salmon, on the other hand, deposits, say, ten thousand eggs, but they are not cared for by the parents, and, therefore, only ten of these, according to Buckland and others, arrive at a state of maturity. I daresay this is not far from the truth. A salmon does not necessarily spawn every year.

We find, then, that the rate of increase amongst salmon and amongst some birds is not so dissimilar as might at first sight be supposed. The case, indeed, seems narrowed down to this, that a pair of salmon in one year add five pairs only to the stock. If the above estimate be correct, and I think it is rather above the mark on some rivers, it is clear that there is a tendency to increase, and we know that in some parts of the world, as, for instance, in the rivers of Alaska, the salmon have increased so enormously that some streams are at times almost blocked by them.

We do not know exactly what is the cause of this excessive increase, nor how long it has been going on, but apparently something has happened that has interfered with Nature's balance (I shall, however, have more to say about these Alaskan rivers presently). Some check has evidently been taken away that has allowed the salmon to increase as they have done. It is, I think, more than probable that the excessive destruction of seals that has been taking place during late years has much to do with it.

Whilst we find, then, that in one part of the world salmon have been increasing, we find that in another part they have decreased to a considerable extent. The cause of the decrease is, I think, too obvious to need explanation. Increased means of capture and number of captors, along with many more indirect causes, have brought about the result. The destruction has been greater than the supply, and the consequence is inevitable. The remedy is also plain—increase the supply by caring for the helpless salmon during its infancy.

A great deal has been said from time to time about salmon not feeding in our rivers. That they feed when in the rivers is beyond any doubt, and there are many instances recorded in which food has been found in their stomachs. On the other hand, there is no doubt that they often fast for long seasons or during those times take but very little food. They are detained in the fresh water longer than they used to be, owing to the low state of our rivers during lengthened periods, due to our drainage systems. That this detention in the fresh water is injurious to the fish is beyond any doubt, and it is difficult to estimate the influence which it may have in the course of a few generations.

Salmon take hardly any food at spawning time or during very cold weather, and when they are feeling rather upset, or " out of sorts," which I believe is often more or less the case when in some of our fresh waters. An idea seems to exist in many minds that the huge bodies of the salmon are somehow developed by a very indefinite something, which the fish manages somehow or other to obtain, by a process which they call " suction." What this " suction " consists of I find they usually do not know. But, in any case, it seems to point to something very much resembling microscopic supplies.

We know that the food of salmon in the sea consists largely of herrings, sand eels, crustaceans, etc. The exceptional and temporary absence of herrings from a portion of our coast has been accompanied by a corresponding scarcity of salmon. The salmon, indeed, feeds voraciously in the sea, and whilst there lives on the very best of food.

During the sojourn in the salt water it has a great deal to do, for it has to recover from the sorry plight in which it often finds tself on its return from the river, and it has also to accumulate a sufficient amount of fat to be of service to it during the time it remains in the fresh water on its next visit. In going to the sea, salmon get a thorough change of water as regards specific gravity, temperature, composition, and products. The return of the fish to the rivers is an exceedingly important economic question, some rivers being early and some being late, and some, no wonder that it should be so, are getting later.

A great desire has been expressed from time to time to make a late river earlier. The way to do this, if it can be done at all, is to deal with the river as well as with the fish. The influences which affect the habits and migrations of the salmon must be carefully considered before attempting to deal with so difficult a problem, but I would by no means discourage attempts to make improvements in this direction. I have seen quite enough myself to be led to the belief that it is quite possible to improve some of our rivers that formerly were earlier than they are at present.

There are two main influences which more or less affect the migrations of all fishes, viz., food supply and reproduction of species. These are the primary elements to deal with in working out any questions bearing on the improvement of fisheries. We know that these items have a very important influence on their welfare, and we know also that there is very much to learn concerning them. In some cases the more we work out the variety of facts connected with the migration of fishes, the more complex do they seem.

We have learned a great deal about the migrations of birds of late years, but, living as they do in the air, we can to a great extent follow their movements, by having observers all over a country or continent, and so we ascertain their exact line of flight. It is not so easy to follow the salmon through all his wanderings, but it is a work that has to be done, and it is not very complimentary to the advanced knowledge of the nineteenth century that we know as little about the "king of fishes" as we do.

We know that the salmon in the "parr" stage, as we find him in our rivers before he has made any acquaintance with the sea, feeds well. We know, also, that about the month of May these

U

little fish lose their trout-like appearance, including the bars, or "finger marks," and that they become bright and silvery. They are then called smolts, and are generally supposed to have got a fresh set of scales, but this is not so. Their shining appearance is caused, not by new scales, but by a silvery pigment secreted on the undersides of the scales. This same silvery appearance also affects the gill covers or opercles, which are not possessed of scales.

An interesting experiment was tried some years ago. A number of salmon "parr" were taken from the river, and placed in a fresh water aquarium tank. In due course about half of them assumed the smolt stage, but the others did not. Sea water was then added until the other was displaced, and, on the water becoming salt as the sea itself, a very interesting result occurred. It has been asserted that "parrs," as such, will not live in salt water. These not only lived, but very soon assumed the smolt stage. We know that salmon "parrs" and smolts when in a river feed voraciously, for they will take nearly any bait. They are found gorged with shellfish, larvæ of insects, etc. So voracious are they that we know full well that they often spoil the sport in a river. Can it be supposed, for a moment, that a salmon in these early stages, during which periods of its existence it does not make any great growth, feeds voraciously, and, after going to the sea, takes either no food or very little, notwithstanding that the "smolt" which leaves the river weighing only a very few ounces, returns very soon as a grilse of several pounds? It seems unreasonable.

Some smolts return from the sea as grilse in about three months, whilst others of the same brood remain in the sea for about fourteen or fifteen months. Those which return in about three months have, taking a very low estimate, reached a weight of some three pounds, but those which remain for fourteen or fifteen months do not necessarily attain a very much greater size than those which return in the shorter period. The same peculiarity is noticeable to an extent amongst trout. Some grow very much more rapidly than others. A smolt let off at Stormontfield in May returned in July of the same year, weighing about three pounds. On the other hand, one which the Duke of Roxburghe let off on May 14th did not return till July of the following year. It had then attained a weight of only six and a half pounds, having in fourteen months just about doubled the weight gained by the other in two months.

During a recent visit to the United States I obtained a mass of information concerning the habits of the various species of salmon inhabiting the waters of North America. By the kindness of Sir Julian Pauncefote, Her Majesty's Ambassador at Washington, to whom I had a letter of introduction, the way was made easy for me to investigate many matters of the deepest interest in connection with the fisheries. As a rule things in America are on a much larger scale than they are here, and this is notably the case with regard to the fisheries. A single catch of salmon, for instance, on some of the Western rivers is sometimes as big as a whole year's catch on a river over here. Seventeen species of salmon are recorded as occurring in Alaska alone, and the largest salmon of the world are credited to that territory.

I am indebted to Dr. Tarleton H. Bean, who is a clever ichthyologist, and holds an important position on the United States Fish Commission, for his generous assistance in gathering particulars respecting the fisheries and fish-cultural operations of the United States. He has personally worked out upon the spot a good deal of what is known of the salmon of Alaska. The fish have been traced as far north as Hotham Inlet, and Dr. Bean says :—" The marine life of the Alaskan salmon is unknown from the time the young, in their newly-acquired silvery dress, leave the fresh-water nursery to become salt-water sailors, until they have ended their cruise, obtained their liberty and come ashore, when, as in the case of so many other salt-water sailors, their serious troubles begin. Salmon remain in fresh water until the second or third spring of their existence, and, not having a bountiful supply of food, they grow very slowly, and seldom exceed eight inches in length when they start seaward. In the ocean they feed on the capeling, the herring, and a small needle-shaped fish called the lant.

" As a rule, the fish remain at sea until they are about to deposit their eggs, and then approach the coast in great masses. A few young males accompany the schools every year, and may or may not return to sea without entering the rivers. The adult

fish come up from the sea at a certain time of year. . . . . .
The length of their stay at the river mouths before ascending, and
the rate of ascent to the spawning grounds depends upon the
urgency of the breeding condition.   In the long rivers the king
salmon (*Oncorhynchus chouica*) travels from twenty to forty miles
a day; this species and the red salmon (*Oncorhynchus nerka*)
are reported to be the greatest travellers.   The silver salmon
(*Oncorhynchus kisutch*) and dog salmon (*Oncorhynchus keta*)
however, are recorded by Dr. Ball as traversing the Yukon at least
a thousand miles.

   "From the time the salmon enters fresh water it begins to
deteriorate in flesh and undergoes remarkable changes in form
and colour.   Arriving as a shapely fish, clad in shining silvery
scales, and with its flesh pink or red, it plays around for a little
while between salt water and fresh, and then begins its long fast
and its wearisome journey.   No food is taken, and there are
shoals, rapids, and sometimes cataracts to surmount; but the
salmon falters not, nor can it be prevented from accomplishing
its mission by anything but death or an impassable barrier.   Its
body soon becomes thin and lacerated, and its fins are worn to
shreds, by contact with the sharp rocks.   In the males a great
lump is developed on the back behind the head, and the jaws
are lengthened and distorted, so that the mouth cannot be closed.
The wounded fish are soon attacked by the salmon· fungus, and
progress from bad to worse, until they become unsightly.   In the
meantime the body colours will have varied from dark grey in the
humpback (*Oncorhynchus gorbuscha*), with the lower parts milky
white to a brilliant vermillion in the red salmon, contrasting
beautifully with the rich olive green of its head.

   " The excessive mortality of salmon during the ascent of the
streams, and on the breeding grounds, has led to the belief that
none of the spawning fish leave the fresh water alive.   There is a
substantial basis for this view in the long rivers, and it is doubtless true that a journey of five hundred miles or more is followed
by the death of all the salmon concerned in it.

   " The silver salmon does not usually ascend streams to a
great distance, and I have seen it return to salt water alive after
spawning.   The red salmon spawns around the shores of deep

cool lakes, and in their tributaries, preferring waters whose highest temperature rarely exceeds fifty-five degrees.

"The king salmon is the first to arrive on the shores in the spring. It makes its appearance in May . . . . and early in June. The time of its coming into Norton Sound corresponds with the breaking up and disappearance of the ice. It continues to enter some of the rivers for the purpose of spawning until August. The height of the season, however, is reached by the middle of July in most localities. This fish travels up the rivers farther than any other species, except the red salmon. In the Yukon it ascends far above Fort Yukon, more than fifteen hundred miles from the mouth of the river.

"The king salmon does not ascend rivers rapidly, unless the spawning season is close at hand. It generally plays around for a few days, or even a couple of weeks, near the river limit of tide water. As far as we can learn, only those fish that ascend the stream short distances return to the ocean after spawning, and September is the month in which the spent fish go down to the sea. There is no reason why the king salmon should not return down the Karluk, as the distance is very short. There is ample testimony, of a conclusive nature, to the effect that, after a king salmon ascends five hundred miles from the sea, it never returns to it alive. The humpback salmon (*Oncorhynchus gorbuscha*) is the smallest, most abundant, and most widely distributed of the Alaskan salmon. The height of the spawning season in the Kadiak streams is evidently about the middle of August. Messrs. Robert Lewis and Livingstone Stone found the humpbacks spawning in vast numbers August 15th. On the 24th of August Alexander Creek was full of humpbacks, in all stages of emaciation and decay. In Alitak Bay, September 9th, the fish were nearly all dead in the creeks, and Snug Harbour contained many dying humpback salmon, floating seaward tail first.

"After the great run in the Karluk, the fish came down dead, or in a dying condition, for a whole month, and the beaches were strewn with red salmon. The last stages of this species are repulsive to look upon, but before the extensive emaciation and sloughing away of the skin has taken place, the colours of the breeding fish are rather pleasing, the lower parts becoming milky white, contrasting beautifully with the darker colour of the sides and back.

" Like the king salmon, the red salmon travels long distances up the rivers, pushing on to their sources ; but it is chiefly a lake spawner, while the king salmon prefers the head waters of the principal rivers to their small tributaries. It is asserted by Mr. Hirsch and others, who have had much experience with the red salmon, that no spawning fish of this species ever leave Karluk river alive. Natives say that they can catch salmon any time uring the winter months, through the ice, on Karluk river and ake. They assert also that all the red salmon die in the spring, mostly in April. It is said that this species will not enter a river which does not arise from a lake.

" In Karluk lake, near the sources of the river, ripe red salmon were speared by the natives, August 17th. On the 18th of the same month large numbers of dead salmon of this species, and plenty of both sexes which were spent and nearly dead, were found in the rivers connecting Karluk lake with its tributary lakes. In all of the little streams falling into Karluk lake in which red salmon were found, dead fish were moderately common, and there was an abundance of young salmon about one and a half inches long, which must have been hatched from eggs deposited during the preceding fall. The male red salmon develops a lump nearly as large as that of the humpback, and its jaws are exceed-ingly enlarged."

These interesting observations upon the salmon of Alaska have much instruction in them. Although there are other species of salmon than our *Salmo salar*, yet they are all salmon, closely allied, belonging to the genus *salmo*, and having a great deal in common, although in appearance, in structure, and to some extent in habit, variations more or less modified occur. When better known, several species may be found to merge into one, but it would be premature for me to make anything approaching a definite assertion on this point at present.

It is very evident that there is a great deal that is common to the salmon of both continents. First of all with regard to the early or fresh-water stage, the young, or "parrs" as we should call them, remain in fresh water until the second and third spring of their existence. That this applies to the " parrs " in some at least of our waters is beyond doubt. There is a good deal of rather confusing evidence forthcoming on the subject, but I have found from experience that some reach the smolt stage the first season, that is as yearlings; some the second, or as two-year-olds; and a few not till the third season, or as three-year-olds. The same occurs amongst trout and char. From the time they enter the sea until they re-appear in our rivers, nothing very definite seems to be known of their habits. This part of the life history of the salmon should by all means be worked out. From the time he enters a river to the time he returns, either to the sea or to

"The dust from whence he sprung,"

we do know something, but we ought to know a good deal more. We know too well that violent fungus epidemics occur in our rivers, and we know also that the greatest destruction of salmon by this means has been followed by great plenty. This is not suggestive of the salmon being a "total abstainer," as so many of his friends would make him out to be. There is no doubt that old kelts destroy a great many young fish, and when these old kelts are removed the result is benefit to the river. The supply in the rivers of Alaska keeps up, notwithstanding the great destruction of adult salmon. One fact is worth noting. The fungus chiefly attacks them after spawning. They succeed then in performing the duty that impels them to thread their way hundreds of miles up rivers full of difficulties and dangers.

In this country, although there are exceptions, possibly owing partly to climatic variations, and to the effects of hill drainage, etc., yet I have found that as a rule the salmon succeed in depositing their ova before they die. I venture the assertion that there is a very great probability that if all the large salmon in a river deposited their eggs and then died, that it might be the best thing for that river that ever happened to it. I do not say that it would be so, but that no great harm would accrue seems to be certain. The salmon of Kamschatka have had their numbers decimated, and yet they have continued their race, and the same applies to those of Alaska.

These facts in the life history of the salmon are instructive.

The further the fish go up rivers of any magnitude the worse it seems to be for them. Some races or species seem to go up the rivers only a short distance, whilst others go rather further, and others again will go more than a thousand miles if they get the chance. One thing is very certain—" From the time the salmon enters fresh water it begins to deteriorate in flesh, and undergoes changes in form and colour." There is a marked difference between a newly-run fish and one that has been in the fresh water for some time. There is one fact that is specially worth noting, and that is that when the fish running up these large rivers enter a tributary, and soon find it blocked up by an impassable fall, they will there deposit their ova, whereas had they held on their course up the main stream, they would probably have travelled hundreds of miles further.

When salmon have done spawning, the sooner they get away to the sea the better. Detention in the river is evidently very bad for them, and they are at such times predisposed to attacks of fungus (*Saprolegnia*). As sea water is fatal to it, those fish which succeed in reaching the sea before they are attacked are freed from further risks, whilst those which have the fungus actually on them are cured in such cases as have not gone too far. *Saprolegnia* is somewhat similar to the fungus of diphtheria, and that is very amenable to medical treatment, if taken in time. On the other hand, in cases which have gone too far, or that have been neglected, death usually results. The same applies to *Saprolegnia*—the treatment is the same, the result is the same. On a fish farm it is usually plain to an expert at a glance whether a fish, if properly treated, will live or die. Every fish-culturist has cases occasionally of *Saprolegnia*. It is one of the commonest diseases among fishes under certain favourable conditions, and it is much better understood than it was a few years ago. As I have already alluded to it in my chapter on pond life, it would be out of place, however, to go further into the matter here.

I regret to hear that some of the American salmon rivers have recently been depopulated by the pollution of the water by the lumber companies. Enormous quantities of sawdust being sent down some of the streams the salmon have been absolutely driven out of them.

APPENDIX.

HOW TO MAKE A RAT TRAP.

ON page 195 I promised to describe a rat trap. It may be called a rat exterminator, as it not only clears them but keeps them clear. Build four walls of stone, brick, or concrete, so as to make a convenient room in which to store the fish food, meat, mussels, etc. Put a coping on the top of the wall so that rats cannot get over it. It may be covered by a roof or not as may be convenient. Concrete the floor. Cover the bottom of the door with sheet iron, and leave a hole in it, or anywhere else where the rats can conveniently enter. Now fit an iron plate to slide in a groove so as to block this hole at will, and attach a wire to communicate with some convenient point, say the keeper's house. Every night let him drop the iron plate and go to bed. The rats will be quite safe till morning. They should have a bundle of straw to hide in, and a terrier will make short work of them as they are turned out. I have known fifty rats taken in one night in such a trap, and it is always effective.

INDEX.